Practical IP and Telecom for Broadcast Engineering and Operations

FOCAL PRESS MEDIA TECHNOLOGY PROFESSIONAL SERIES

Coming Soon:

Video Over IP: A Practical Guide to Technology and Applications, by Wes Simpson, Telecom Product Consulting

The HDTV Handbook, by Randy Hoffner, ABC

Entertainment Content Protection: Technology, Rights Management, and Legal Constraints, by Scott Hamilton, Fox Entertainment Group

Practical IP and Telecom for Broadcast Engineering and Operations

Fred Huffman

Focal Press
Taylor & Francis Group

NEW YORK AND LONDON

First published 2004

This edition published 2013
by Focal Press
70 Blanchard Road, Suite 402, Burlington, MA 01803

Simultaneously published in the UK
by Focal Press
2 Park Square, Milton Park, Abingdon, Oxon OX14 4RN

Focal Press is an imprint of the Taylor & Francis Group, an informa business

Copyright © 2004 Fred Huffman.

Library of Congress Cataloging-in-Publication Data
Huffman, Fred.
   Practical IP and telecom for broadcast engineering and operations /
   Fred Huffman. — 1st ed.
      p. cm.
      ISBN 0-240-80589-5
      1. TCP/IP (Computer network protocol) 2. Internet. 3. Telecommunication.
   I. Title.
      TK5105.585.H84 2004
      621.382'12—dc22

                                                                  2004002907

British Library Cataloguing-in-Publication Data
A catalogue record for this book is available from the British Library.

ISBN 13: 978-0-240-80589-4 (pbk)

TABLE OF CONTENTS

INTRODUCTION

This book is about knowledge. It is not another treatise on all you need to know about Broadcasting, Internet, or Telecom. The focus is on what you need to know about using Telecom and Internet equipment, facilities, and services to transport content in broadcasting or similar operations. It is not a biased pitch for classical terrestrial telecom networks or satellite networks. Its target audience includes executives, managers, and professional staff engaged in engineering, on-air, and content production operations. The book is in two parts. The first is technical in nature, while the second is business oriented or, as one of my reviewers said, "administrative."

Practical Internet and Telecom for Broadcast Engineering and Operations gestated in the fall of 2002 following the "Streaming For Broadcast Engineers" tutorial sessions for the IEEE Broadcast Technical Society. The first seminar was conducted in October 2001. Attendees were mostly seasoned broadcast engineers with many years experience. The second seminar occurred in April 2002 and attracted a remarkably larger group with mixed backgrounds. No formal survey was taken, but it was easy to discern the group consisted of broadcasters interested in learning about Internet technology and Internet engineers wanting to know more about video.

Preparing the lecture material, presenting it, and working through the sessions with questions and dialog brought clarity to perceptions that were previously vague at best. Two observations were made:

- Broadcast engineers could benefit from learning about transporting content over Telecom and Internet facilities, and
- Internet counterparts could benefit from learning more about the nature and characteristics of content. Maybe there's a second book in the future. For present purposes though, the scope and focus is broadcast engineering and operations professionals interested in learning about Internet and Telecom technology.

The aim of this book is to explain how telecom practices, products, services, and technology can be used effectively in broadcast engineering and operations, where *operations* is taken in the broader context of the overall business as defined by enterprise mission and financial statements. Use of the term *broadcast* is figurative and historical. Other professionals will find the publication useful in conferencing, education, training, or similar use of audio and video signals.

What does the practicing media industry professional need to know about Internet and Telecom? This book will enable you to learn business and technical fundamentals to support your day-to-day work and help you survive over the long-term. The book is structured in two parts. The first part is technical at a network or system level. Around Chapter 6, it begins to move from a purely technical vein into defining applications, specifying equipment and services, budgeting, and long-term planning. Essentially, the second part addresses how to take the technology and fit it into business.

- Current and emerging digital communications network architecture
- Clarifying telephone, Internet, multimedia jargon and lingo
- Analyzing and understanding Internet and telecom facilities
- The characteristics of digitized content from a telecom perspective

- How to relate, map, and interface digitized content to digital networks
- How to design, specify, and get internet/telecom services at affordable cost
- Developing strategic network plans that fit enterprise financial models
- Monitoring regulatory and technology trends for long-term survival

Why should practicing broadcast professionals know about Internet and Telecom?

- Life in the real world is competitive. Basic human tendency is to survive, grow, and prosper. For most, the alternative isn't very attractive.
- Change is inevitable. You can ignore or fight it. You might find it more palatable to embrace it, learn from it and *survive, grow,* and *prosper.*
- If you choose to embrace change rather than react to it, you may find you'll live longer and get a raise and a promotion or two. Move that file you just created or an old one from the archives with your mouse instead of calling a courier or overnight delivery service; it's faster and more secure. Picture using a home theater remote control to rent or buy the latest Hollywood release instead of driving around the corner or across town. Better yet, have your home theater search or watch for certain content and when it finds it, capture it, and notify you that it's available when you want to take time to "tune in." Sure, accomplishing all these great things is quite possible with other media. However, "network" media may well give you a cost effective, competitive edge.

The level of detail and amount of information about audio, video and other aspects of broadcast and technical operations may strike some as more than necessary or appropriate. However, nothing in the book is intended to substitute for professional broadcast knowledge and expertise. The intent is to dovetail and converge Broadcast, Internet, and Telecom expertise. Perhaps bridge your knowledge from one field to another, or between all three.

KEY CONCEPTS AND MODELS

The overall approach to the subject relies on the following three models: a "Program Content Food Chain," a Network Interface Device (NID), and a block diagram depicting End-to-End Service. All three models are explained in detail in Chapter 1. Because of their importance though, they are briefly described here as part of the introduction.

Figure 1 illustrates the concept of what we call the Program Content Food Chain. Simply a diagram to show three phases most content goes through during its life. Content creation, distribution, and delivery require a vehicle to carry it through all three stages of its life span. These stages are defined as *Creation, Distribution,* and *Delivery.*

Around the time the industry standardized digital studio systems the terms *Contribution* and *Distribution* appeared because a way to distinguish digitized content from analog transmission was needed. This two-stage model works quite well with analog transmission to the end user or viewer. The Program Content Food Chain model is a further extension of the two-stage model and was conceived to recognize, quantify, and leverage the ability of digital transmission systems to deliver multiple levels of picture quality and sound fidelity to the end user or viewer.

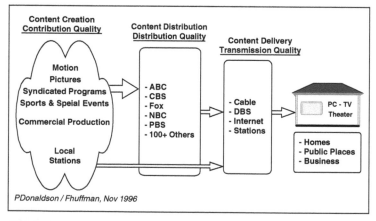

FIGURE 1   The Program Content Food Chain Illustrated

Widespread use of lossy compression techniques to reduce transport bandwidth results in impairment of picture quality and sound fidelity as content travels the food chain path. In addition to recognizing the process of creating, distributing, and delivering content, the food chain model provides a way to characterize each phase with a quality metric. Establishing a quality classification allows practical sizing and selection of network bandwidth and cost.

The second model is a Network Interface Device (NID). The NID functions as a standardized interface between broadcast and network facilities. Another way to view the NID is to see it as the enabler of network based content transport between production, distribution, and delivery phases of the program content food chain. Figure 2 is a block functional block diagram of the concept of a network interface device.

Realization of a practical network interface is likely to require use of more than one commercial off the shelf product. In today's environment where constant churn and change are a fact of life, multiple products and a constant insistence on compliance with formal, recognized standards provide the ability to migrate from one generation to the next as an alternative to wholesale replacement of single items of equipment and software.

The main message in the NID concept is standardization, where Broadcast Standards are invoked on one side; Internet and Telecom Standards on the other.

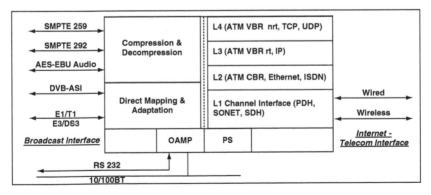

FIGURE 2 Network Interface Device Block Diagram

The third model we want to use is an End-to-End Service Model. The notion behind such a model is to recognize another fact of life in networking and that is the basic connection or movement of information through a transmission facility. Even though we think of broadcasting and multicasting, audiences and their source of revenue are made up of individual viewers. As far as the viewer is concerned, there is only one path and that's between their set and where ever the program content is coming from. Move back through the program content food chain, and it's the same for the broadcaster. What's important is the connection between their transmitter input and where the content is originating. As long as it arrives and goes out intact, not much else matters.

Moving content through networks involves three components: the NID, Network Access, and Network Transport. Figure 3 shows the basic End-to-End Service reference model.

One final notion about the nature of the end-to-end service model is the fact that it can be applied to sub-nets and individual links in an overall network architecture. Designing, specifying, accepting deliverables, or troubleshooting on a design or component failure level can be conducted on individual components delivered by one or more suppliers in a straight forward, scalable, objective fashion, eliminating finger-pointing.

These three components work well in single, point-to-point links without network management; however, the ability to change various configuration parameters and provide performance parameters and report errors or failures enhances operational success. As the number of locations served increase, network management becomes a critical, necessary function.

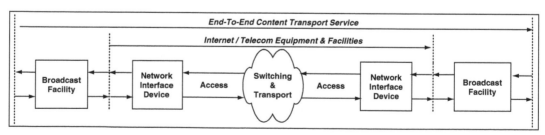

FIGURE 3   Block Diagram End-To-End Service Components

These three models are generic and can be used as a basic foundation for day day-to-day design, planning, and operations. There's very little about the subject matter in the book that doesn't have a place somewhere in one or all the models; there's not much you need to worry about that one or all three do not cover.

SURVIVING LONG TERM

How to survive long term is an underlying philosophy throughout the book. Someone once said, "Hand a man a fish and feed him for a day. Teach him to fish and feed him for life." This book is in large part, a reflection of experience and knowledge accumulated over the course of 30+ years conceiving, designing, proposing, selling, building, operating, breaking, and fixing equipment, software, systems, and networks. This experience is about evenly divided between broadcast and communications systems.

In December 1982, after 15 years at RCA's Broadcast Systems Division, I transferred to a new job in RCA's Communications Group. In the same month, AT&T and the US Department of Justice entered into a consent decree that fractured the company into independent Local Exchange and Inter-Exchange Carrier companies. The era of Ma Bell would be no more. Over the next few years with RCA and a small business I owned, I designed and built corporate networks with local and long distance voice and high-speed data communications capability. I was fortunate to work with some of the early voice processing systems with auto-attendant, data collection, and voice mail networking capabilities.

MCI hired me to work in its International Network Engineering Group in 1991. My job was to design and build a mainframe based computer tool that could be used by finance, engineering, and marketing to plan and manage the company's investments and interest in undersea fiber cable systems. I knew very little about mainframe computer systems or database management software, but I knew digital communications networks very well, had significant knowledge of international commerce, and was quite willing to learn about data base manger technology.

During the course of the next $11\frac{1}{2}$ years, I was fortunate to work with a range of technologies and systems reflective of the convergence of content and communications networks in the emerging Internet environment.

Working through this period gave me opportunities to learn from experts with knowledge of both fields in a real hands-on experience. This learning experience continually crossed between relating what I knew as well as digging in and learning from others who knew more about a subject of interest than I did.

Looking back now it's pretty easy to see that radio and television—"broadcasting" as we know it—was undergoing massive change driven by technology and gated by regulatory actions. "Moore's Law" was making an impact.

I'm grateful for many gifts over the years. High on the list is a strong interest and appreciation for learning. I strongly believe my interest is a direct result of the influence of parents and other family members interest in reading and learning. Call it a "desire to know"—maybe curiosity—fed by teachers, librarians, writers, and more recently, the wonders of the Internet. This basic trait has served me well over the years and continues to keep me going. Associated with this characteristic is the vast sea of knowledge that seems to keep growing almost faster than any human can hope to keep up with.

With this I wish you well and hope this effort contributes to your curiosity, sense of learning, and ultimate success in your professional and personal endeavours.

Fred W. Huffman
Richardson, Texas
March 6, 2003

1

BASIC FUNDAMENTALS, DEFINITIONS, AND KEY TERMS

Picture a group of 104 people in a room at a conference. The make up of the group is about 51% telecom heads, 45% transmission equipment manufacturers, and 5% broadcasters. The broadcasters finished their once a year presentations, yet another attempt to tell the telecom people that their network performance is and has been unsatisfactory, again (!), for the past year. One more time they didn't get it. A broadcaster cites a 3-hour service interruption and states emphatically and profoundly that they couldn't even get a telephone call acknowledging the outage until an hour after service was restored. The amount of time, energy, and passion put into attempts to gain a common understanding of *service-outage* would amaze and amuse even the most disinterested casual observer.

What's the problem? In a nutshell, and no pun intended, the group is suffering from a communications problem. Actually, that's not the problem. It's a symptom of more deep-seated malfunction in the practice of human behavior. Any reasonably objective and not necessarily disinterested observer would see there are several issues in the situation described. At the top of the list of issues is this simple fact: these good people weren't listening to each other. Therefore, they couldn't possibly absorb what was being said on the other side and had no chance at reaching a common understanding of and resolution of their age-old common problem(s). Well, maybe they tried, but it was an unsuccessful, feeble attempt at best, to define and reach a common understanding of the term outage. Far too often lack of progress in such encounters is simply because the participants use words and phrases that have different meanings to each other, or they use different words and phrases to mean the same thing and don't realize it.

This chapter begins to solve problems and challenges dealing with Internet and Telecom facilities, networks, and services such as those mentioned previously. It introduces key concepts, defines basic terms, and suggests a process that will enable the reader to organize and successfully manage an organization's assets and expenditures for communications equipment and services. It covers the following major topics:

- Definition of content and content transport
- The program content food chain
- The migration from analog to digital content transport
- Description of the functions of a network interface device
- Commonly misused and misunderstood acronyms and terms
- A survey of current and applicable standards

From the perspective of dealing with communications network subject matter, you will find it advantageous to limit what you say or write, and be precise and clear about it. The most critical parts of your work with internal information technology (IT)/Telecom experts and third-party suppliers can be found in a few topics, such as bitrate, interface, class of service, and network performance. If you don't understand it and can't communicate it, you risk being misunderstood or, worse yet, mislead or taken advantage of. On the other hand, if you understand and can explain it, you can question and make sure your suppliers and colleagues understand. They might not agree, but that's another story.

The material in this chapter lays a foundation for the remainder of the book. The three models, mentioned in the introduction—Program

Content Food Chain, Network Interface Device (NID), and End-to-End Service—are defined and explained. All are current and relatively undated. However, it would be unrealistic not to expect features and functions of the NID and end-to-end service model to evolve over time. The food chain model should be stable and useful for a long time.

CONTENT AND CONTENT TRANSPORT DEFINED

Content, in simple terms, is information in the form of audio, video, still and moving images. Information from transducers, which monitor anything from pressure in a pipe to earthquake vibrations displayed on a Richter scale, could be deemed content. Content is sometimes confused with media. In the event confusion occurs, sort it out by thinking of the paper cup and string model. The sound traveling from one cup to the other is content and the string between the two serves as media. Think of the cups as the interface between the content and the media.

Audio and video signals begin life in analog form. Audio is aperiodic, unstructured, and reflective of the dynamic, random nature of sound. Once converted to digital, it takes on a rigid, blocky, periodic structure. Video is different. It is highly structured, reflecting the nature and character of a scan mechanism. The scanning mechanism supports organization of a frame of information coming from a relatively large group of individual sensors, mechanically and electrically sectioned to capture small parts or areas of a scene. With some amplification and care, the image(s) can convince humans that they can see something in the way of a reasonable and acceptable representation—*analog* of the real scene.

Digitized audio and video content become an *object* in computer lingo, capable of being automated and manipulated in many ways not remotely possible in analog form. Content in bits, bytes, cells, or packets can be encapsulated in containers and interfaced to digital transport media. Clocking, synchronization, and timing become critical parameters and characteristics that must be constantly attended to.

Content transport means to move the content from one point to another or from a single location or source to multiple locations.

Content can be transported live, or in *real time,* and used immediately or stored for further editing and packaging before being staged for the ultimate purpose.

An often-used metaphor has digitized content in liquid form that is carried in a digital pipe. Supposedly, the content flows through the pipe like water. Not exactly an ideal description, but close. Maybe a more appropriate metaphor would use marbles instead of water. Imagine replacing marbles with watermelons. Both have unique characteristics, are quite different in size, and respond differently to the relatively fixed and rigid characteristics of any particular transport media type. With a little more imagination, it is easy to see that both should be packed and protected differently for the trip. Packing marbles with material normally used to protect watermelons may turn the marbles into small pieces of glass by the end of the journey. On the other hand, packing watermelons in material suitable for transporting marbles may make watermelons unaffordable at the point of sale.

In summary, and to be practical, content is valuable property as long as it gets to the point of a transaction. Someone has to want it, and order it. Once that's done, it must be delivered. Content transport simply delivers the valuable property. Somewhere in the process, money or other good and valuable property passes in the opposite direction.

THE PROGRAM CONTENT FOOD CHAIN

The program content food chain envisions three distinct stages: creation, distribution, and delivery. Each stage is unique. Designing and building networks to transport the content within and between each of the stages must address the unique nature and quality characteristics of digitized content. For reasons presented later, the nature of the content is quite different in each of the three stages. It's not a "one size fits all" world anymore. The days of one analog program on one analog transmission channel are numbered.

The program content food chain model illustrates the flow of content from start to finish. The path of travel for content was initially, and still is in large part, made up of analog vehicles. Moving content in this kind of environment is very limited, and subject to noise, distortion, and deterioration found in passive conductor and radio wave transmission facilities. Initially and for many years, the life cycle of content was very short. Essentially limited to real-time or live sessions only, the invention of recording and playback schemes enabled multiple use of content. This basic capability enabled non–real-time content transport by physically moving the media it was stored on. For the purposes of this book, we will focus on network media, not storage.

Gradually, content production equipment and systems have migrated from analog to digital technology, driven in large part by less expensive electronic components developed for the computer and telecom industries. In fact complete, all digital, end-to-end paths are possible with cable modems, terrestrial, and satellite transmission facilities.

Figure 1-1 is an illustration of the three stages of the life cycle of program content. It also shows the relative quality of each stage and paves the way for later explanations of how picture quality and sound fidelity can be matched to transmission bandwidth and cost of transporting content throughout the process of creating, distributing, and delivering content.

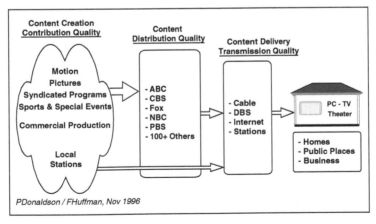

FIGURE 1-1 Program Content Food Chain

The first stage in the chain is content creation. In this stage, raw audio and/or video is captured, or created. Still images or scenes can be captured on film, printed on paper or, if captured with a digital camera, transferred to a computer system and ultimately added into a program or a piece of motion picture entertainment. Titles and other graphics can be created on computer systems and included in the final package.

In terms of picture quality and sound fidelity, this is as good as it gets—"the best." Compression and decompression may or may not be used in content creation. Contribution quality is a term that's been used for many years and is likely to remain a well-understood metric long into the future.

Content distribution is an intermediate state. In the past, and still to a great degree, distribution of content is understood to encompass all processes between creation of the content and use by the viewer. This is quite sufficient for analog content, but is unsatisfactory for digitized content. Digitized content enables multiple versions and repurposing master material across multiple applications. It can also be packaged to fit the myriad of delivery vehicles, all digital, that currently exist and are likely to continue to evolve into the foreseeable future.

Much like physical goods distribution, the transport process requires multiple vehicles and transfer points depending on volume. It's likely the content may be handed off in compressed form. If it's not handed off in compressed form, the first step after hand-off is probably compression. At this point, the content is either stored or transmitted. Sometimes both happen. It's almost certainly subject to degradation because of the effects of lossy compression.

Distribution quality as a term has been around for many years as well, and has been used to characterize quality in what is defined in the food chain model as two separate and entirely different stages.

Content delivery is the third and final stage in the food chain model. This is the point where content is handed off to the end user or target audience. It's almost guaranteed to be in compressed form and must be decompressed before being used. This is true for live, real-time transport or in a file transfer or non–real time. This is the point where it's decompressed the last time before it's used or transmitted to the viewer or target audience, thus the label *transmission quality*.

Why the food chain model? It is important to recognize some basics. First, program content will have experienced compression-decompression at least once, perhaps several times as it moves through the food chain. Second, the compression process is at best a mix of lossless and lossy, or more likely, all lossy. Lossy compression simply means information bits are thrown away in the compression process, and unrecoverable in the decompression process. For example, a given group of bits making up all the bytes in a television (TV) frame won't be the same when it's decompressed. Lossless compression on the other hand is a controlled and intentional process. It ensures bit-for-bit replication of the original source material, each and every time the compression and decompression process takes place.

Picture quality and sound fidelity of the delivered content are directly related to, and dependent on, the bit rate or payload of the compressed signal. Higher bit rate payloads result from less compression and its effect on picture quality and sound fidelity.

Another consideration is amount of detail in the source material. For example, digitized source material can be achieved with ITU-R BT601, and contains far more detail than the same material after it has been encoded for National Television Systems Committee (NTSC), Phase Alternate Line (PAL), or Sequentiel Couleur Avec Memorie (SECAM) transmission. These processes include filtering to limit the amount of high frequency detail in order to reduce the level of interference with the color subcarrier. It makes little or no sense to compress analog signals into payloads greater than approximately 10 Mbs because there is very little or no detail left in the video signal after it has been filtered and passed through the encoding process. If the detailed picture information doesn't exist, it can't be compressed and passed on. Higher bitrate sampling and encoding results in unnecessarily higher bandwidth (more bits) in the communications channel (unnecessary cost) and any noise or artifacts generated degrades picture quality.

On the other hand, compress the same material beyond approximately 8 Mbs and quality goes to pot quickly, as evident from loss of detail and increase in noise, and artifacts will show. Not coincidentally, many broadcasters and cable programmers have adopted 8 Mbs payload as their standard for content delivery applications, which seems to be a reasonable digital equivalent to analog *broadcast quality*.

Yet another side of this complex picture is the possibility of realizing equivalent quality from compressing BT601 digital material that has not been subjected to NTSC, PAL, or SECAM encoding, to 4 to 6 Mbs payloads. Results will vary because of different approaches to product implementation and system configuration.

Facing the reality that picture quality and sound fidelity take a hit and accumulate as compression and de-compression cycles occur, the issue becomes how to deal with it on an end-to-end basis. The general approach is to define and quantify the level of picture quality and sound fidelity required at delivery. Once determined, this can drive compression configuration through the food chain based on a trade-off of quality vs. bandwidth cost. Higher payload bitrates require more transport bandwidth, driving transport media cost higher.

One final point, which is a source of confusion among broadcasters and network people, has to do with variable bitrate program content and variable bitrate transmission channels. First of all, there are two basic encoding techniques. One fixes the amount of bandwidth used and allows picture quality and sound fidelity to suffer. The other maintains quality and allows the amount of bandwidth to be whatever it turns out to be.

On the transmission channel side, time division multiplexing (TDM) channels such as ATSC A-53, or an E3 or DS3 are rock solid, fixed bandwidth transmission channels, sometime called *pipes*. Of the five classes of ATM transmission, only constant bitrate (CBR) is fixed. ATM variable bitrate real time (VBR-rt) may or may not be satisfactory for fixed *or* variable rate program content. Most likely it won't. Here's why: VBR-rt channel characteristics are defined in terms of peak cell rate and sustained cell rate (SCR). Carrying fixed rate content on a VBR-rt channel requires SCR to meet or slightly exceed the payload bitrate. Any less bandwidth results in errors in transmission. Carrying variable rate content in variable rate transmission channels requires the peaks in both to match, that is, the peak transmission channel bandwidth must match and be available when the peak payload requires it, or errors in transmission occur. There is no known mechanism for synchronizing payload content with network transmission under these conditions. Thus the outcome is limited by the statistical probability of network bits available for unknown peak program content payload bits. Therefore, the only way to guarantee error-free transport of variable rate program content is to use CBR service, or if VBR-rt service is chosen, set the Sustained Cell Rate (SCR) of the transmission channel to match the payload bitrate. However, bear in mind that CBR traffic gets a higher priority than lower class variable bitrate traffic when it comes to restoring the network from a path outage. Another undesirable characteristic, especially for live traffic where the content takes the form of an interactive interview, is the additional delay experienced by variable bitrate traffic.

THE MIGRATION FROM ANALOG TO DIGITAL CONTENT TRANSPORT

Consider the traditional analog world in which program content is backhauled with analog telecom facilities and 34-MHz full- or half-transponder satellite transmission facilities, then delivered to radio and television audiences over amplitude modulation (AM), frequency modulation (FM), TV, and cable transmission systems. It's been around since the beginning and works well, one program per channel.

On the other hand, newer digital transmission systems such as Direct Broadcast Satellite (DBS), Digital TV (DTV), digital radio, and cable services use the same radio frequency spectrum structure, but the transmitting and receiving equipment is digital instead of analog.

The DBS architecture permits 32 carriers or transponders in each satellite orbital slot. Each transponder can carry 6 to 10 standard definition programs allocated into timeslots or virtual channels in a 30Mbs digital bit stream occupying approximately 30MHz RF channel bandwidth. Similarly, 6 standard definition programs (maybe more), or 2 high-definition programs can be fitted into virtual channels created from a 19.4Mbs digital bit stream occupying 6MHz RF channel bandwidth, within the Federal Communications Commission's long-standing physical channel allocation plan

AM/FM broadcasters are converting and upgrading existing transmission systems to carry digital bit streams enabling multiple audio programs to be carried within existing RF bandwidth and channel allocations.

Cable systems and telephone companies are upgrading and building new facilities to enable access to the Internet and deliver program content in standardized, digital, virtual channels.

The result of this migration to digital transmission technology is simply more channel capacity capable of distributing and delivering more program material. The total bits in each channel can be subdivided into virtual channels of almost any conceivable amount of bandwidth, as long as the aggregate and multiplexing overhead bits fit within the capacity of the channel. On a business level, digitization of the content delivery mechanism continues to fracture advertising and subscription revenues into smaller and smaller audiences even down to the individual consumer in a concept called *video on demand* (VOD).

THE NATURE AND CHARACTERISTICS OF CONTENT

Analog audio and video signals are usually converted into digital form within physical proximity to the source, ranging from a few inches to tens of feet, perhaps a hundred feet at most. Once in digital form, many things can be done with it, including messing it up unless it's handled with care. Analog content, or content in analog form, requires an analog transmission channel, or it must be converted to digital form for transmission on a digital transmission channel. Digital content requires a digital transmission channel. Analog content by nature must run in real time. Even variable motion recording and playback equipment records the material at normal speed and simply repeats the same frame over several repetitions of vertical synchronizing time base to get the appearance of slower motion, or speeds up several frames over a period of time, giving the viewer the same information faster so it appears in fast motion.

Analog audio played slower or faster than the speed at which it was recorded, sounds speeded up or slowed down with a directly associated change in pitch. The faster the material is played, the higher the pitch. Conversely, the slower the playback speed, the lower its pitch. Digitized audio can be played so that speed of delivery and pitch are independently controllable and variable. Digitized video can have pixels repeated or skipped in a line and it won't be noticed. Lines in a frame can be repeated or skipped; even skipped or repeated frames are hard to detect, especially if they are skipped or repeated within the context of an undisturbed display scan.

Analog content in analog transmission channels or baseband signals on wire are susceptible to myriad distortions and interference. Effects of distortion and interference tend to be incremental and linear. Digital content on the other hand is either resistant to, or immune from interference, or when it fails, it deteriorates rapidly giving rise to a characterization called the *cliff effect*.

In addition to audio and video signals, content may originate in computer systems where it's created by software routines running on hardware. Closed captions appearing at the bottom of a picture are often created by speech-to-text conversion equipment. In any case, and regardless what form the content is in, it's rarely possible to transport it without some further processing, formatting, and multiplexing. Depending on the value of the content, and the number of delivery or reception points, other measures may be required such as encryption, error protection, or replication.

Audio Content

The concept of how a microphone converts sound waves to electrical energy is simple and well known. Audio signals are derived from microphones. *Good sound* requires well-behaved, well-placed microphones. Connecting the analog output of microphones to a small select group of electronic circuits such as analog-to-digital converters, filtering, processing, can produce *good digital sound*. Figure 1-2 is a functional block diagram showing the major elements of a single channel audio source model.

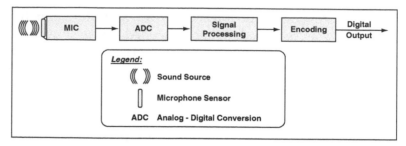

FIGURE 1-2 Audio Source Block Diagram

The sample rate must be high enough to accommodate the analog bandwidth, usually at least twice the rate of the highest frequency in the spectrum of the signal to be digitized. For example, 3.5-kHz toll quality speech is sampled at 8-kHz rate. Professional recording requires frequency response to 20 kHz or more, and relies on 48-kHz sampling rate.

Quantizing parameters are selected and set to faithfully capture and preserve all the information in the analog signal. Quantization parameters for toll quality speech are 8 bits, and professional recording ranges from 16 to 24 bits. Eight-bit toll quality speech has 256 discrete levels across a range of 48 db. Sixteen-bit professional sound permits 1024 discrete values or steps and encompasses a range of 96 db. Increase quantization to 20 bits and the range increases to 120 db.

The output can be saved in a file, or interfaced to digital transport media. Regardless, this entity is best thought of as an "Object." This object can be moved into storage or across a network transport medium, never to be in analog form until it meets speaker or earphone wire and terminals. If desired, the object can be transported with transmission control protocol/Internet protocol (TCP/IP) as a simple file transfer, safe, secure, and reliable like any ordinary document, binary, or other file.

Stereo sound requires a matched pair, including microphones. An early successful example of stereo digital recording was Jack Renner and Bob Woods recording the Cleveland Symphony Orchestra in 1978. Taking the first of many risks characteristic of Telarc's history, they made the first commercial classical recordings in the United States in digital format. One of those recordings was the legendary

Tchaikovsky 1812 Overture, performed by Erich Kunzel and the Cincinnati Pops Orchestra. The recording featured the first-ever digitally recorded live sound effects—digital cannons. Twenty years later, Telarc International released a 6-channel recording of the Tchaikovsky's 1812 Overture, "...complete with new bells, cannons, carillons and choruses."[1] The technical approach was built around 48-kHz sampling and 16-bit quantization using matched microphones and recording channels whereby the sound bits and bytes were captured to magnetic tape, all $48,000 \times 16 \times 2 = 1,536,000$ bits per second. This work would lead standards development by several years, including AES5-1984 r1992, AES Recommended Practice for Professional digital audio, covering preferred sampling frequencies, and AES3-1985 r1992, AES Recommended Practice For Digital Audio Engineering, covering two-channel serial transmission format.

Multi-channel sound applications such as multi-language program audio and multi-channel systems require four, five, or more discrete sound channels. Transporting multiple channel sound requires an interface and a specific set of network resources for each discrete sound channel. The aggregate of the individual channels makes up the total payload offered to the network or recording media transporting the payload.

The output of a microphone, once digitized, is typically fed to an audio console where it is mixed with other microphones and sound sources such as compact discs (CDs), film sound, playback from systems, off-site remote locations, and even test signal generators. These audio consoles vary in complexity in terms of the number of input and output channels, and functional capabilities achievable through manual and software controlled mixing, equalization, delay, synchronization echo, and others. Audio console output fed to a recording or transmission channel is typically a standard interface and signal format. The content is uncompressed and the interface becomes a convenient physical point to connect to a transmission or recording channel.

The commonly used reference in broadcasting and similar applications is AES/EBU. The original standard, AES3-1985 has been modified and updated over the years. Current network design and operations practice revolves around specification of the number of channels and bitrate required through the transmission channel. Typically this is simply stated as AES/EBU two or four channels, and 384- or 768-Kb bandwidth. Five-channel surround sound is becoming more widely used, and practice has become a matter of specifying the source as five-channel surround sound and the amount of bandwidth desired.

Summarizing, audio content is created using digital technology that is standardized around interface, bitrate, quantizing depth, and coding practices. Network facilities capable of supporting audio content transport is a matter of understanding the capabilities and limitations of the chosen standard and matching the nature and characteristics of content package to the requirements placed on the network transmission channel. From a network design and operations perspective, we care about the payload bitrate and interface. For example, if the audio source side of an interface is said to be compliant with AES/EBU, then the network equipment receiving the payload must also be compliant with the same standard. The signal or audio information is un-compressed. If compression is to be used, then the compression equipment between the audio source equipment and network equipment must be compliant with the AES/EBU interface on the source side, matching Internet and/or Telecom interface on the network side.

The Audio Engineering Society (www.aes.org) is the source of AES/EBU standards. The Moving Picture Experts Group (MPEG) is the source of MP3, a compression standard (www.chiariglione.org/mpeg/standards.htm). Another important standard covering digital television sound transmission is A52, ATSC Standard Digital Audio Compression (AC-3; www.atsc.org).

Video Content

Think **compression** is a modern technique? Think again. Monochrome television pictures, or *video*, originated around the notion of a scanning mechanism conceived by two inventors working independently in the early 1900s, neither of whom was aware of the other's idea until many years later. Essentially, the system approved by the FCC and adopted with minor variation by government and regulatory bodies in the rest of the world over time, involves two line rates and two frame rates. Along the way, someone figured out and probably filed an improvement patent on a technique called *interlaced scan*. Interlaced scan cut the required transmission bandwidth in half without significantly impairing perceived picture quality. Essentially, half the picture is traced on the screen one line at a time, and then the process repeats from the top of the picture, scanning additional lines between the lines of the initial area. Thus, fields and frames were born, a field being one-half of a frame. Occasionally the interlaced scan technique is referred to as *2:1 compression*.

Compatible color television was standardized in the United States in 1953 when the NTSC recommended and the FCC adopted rules covering transmission of color television signals. Use of the term compatible means that monochrome receivers could receive and display color signals on a monochrome display from signals transmitted and received on color receivers as a color signal. The reverse is also true, to the extent that color receivers can receive and display monochrome pictures. Several years later, similar practice of backward compatibility followed with adoption of stereo transmission technology in FM radio and television sound.

Color television pictures are made from primary color signals derived from red, green, and blue signals produced by pickup devices, originally scanning electron tubes and, more recently, charge couple device (CCD) solid-state sensors packaged in various forms. Basically, signals from the three sensors are processed to make a monochrome picture that monochrome receivers can detect and display, while additional detection and decoding circuitry in color receivers extract the color information and "paint" it on, or into monochrome pictures. Color television leverages the basics of color derived from combinations of the color primaries (RGB) into three secondary colors—cyan, magenta, and yellow. With appropriate mixing, it is also possible to make black, white, and a range of monochrome tones in between, also called a *grayscale*. Television had its own version of luminance and chrominance. The ability to include color information in the same channel by simply multiplexing the three colors into a signal that enables recovery and representation of the original image has been extended to describing composite color signals as an example of 6:1 compression, 2:1 for interlaced scan, and 3:1 for RGB color into chroma.

Highly summarized, here's what happens. Light reflected from the scene is separated into three primary colors and scanned. Processing the color information and encoding it into a chrominance signal along with a second synchronizing signal permits the color receiver to decode the color information and pass it to the red, blue, and green electron guns in the display tube, which make color pictures by illuminating red, blue, and green phosphor dots on the face of the display tube. Color signals before encoding are referred to as component signals. Color signals after encoding into NTSC, PAL, or SECAM are called composite signals.

Video signals derived from electron tubes or CCD sensors are analog in nature. The signal is highly structured and based on a precision clock or synchronizing signal making up the display scan structure. After transfer from the sensors, it's processed and in the case of the old electron tubes, encoded into a composite color signal. In the case of CCDs, the signal is immediately converted from analog to digital format.

Sometime in the mid-1970s, composite analog television video began to be converted to digital format and processed for several functional purposes. Two of the earliest uses of digitized video included time base correctors in video tape recorders and synchronization of remote, non-synchronous video through devices simply called *frame store synchronizers*. Both devices sampled the analog signal, quantized it into representative samples, and stored the samples in memory, to be read out later in time, synchronous with local station video. In the case of the time base corrector, the storage was a few lines, but the frame synchronizer stored a full frame (525 or 625 lines).

Over time, digital versions of composite video have become **composite digital video** and, as one might expect, many incompatible sets of equipment evolved into the marketplace. Relatively mature versions stabilized around sampling rates equal to four times color sub carrier, commonly termed $4f_{sc}$ and quantized to 8 or 10 bit depth. The industry has evolved to a common serial digital interface, defined in SMPTE 259M for 525/60 and 625/50 scanning formats, making up composite and 4:2:2 component signals. SMPTE 259M is a common interface reference for standard definition television (SDTV) studio mixing, switching, special effects, monitoring, and compression equipment. It covers four payload bitrates, 143Mbs (NTSC), 177-Mb (PAL), 270Mbs 525/625 component (4:3 aspect ratio), and 360Mbs component (16:9 aspect ratio).

SMPTE 292M is a similar interface covering high definition television (HDTV). The maximum payload bitrate is dependent on the source format of the video signal, but doesn't exceed 1.485Gbs. As in SMPTE 259M, SMPTE 292 is commonly used to refer to input and output signals of various mixing, switching, monitoring, and compression equipment. In the case of compression equipment, the SMPTE 292 reference is for the input to the compression equipment. The output standard would be based on the particular recorder or transmission channel standard.

Transmission bandwidth of a single analog video signal varies by application, but is generally many times greater than an accompanying audio signal, even considering stereo or multi-channel sound. Standard definition analog video signals carried on NTSC or PAL transmission facilities are intentionally bandwidth limited to approximately 4 to 5 MHz. This constraint was arrived at as a compromise between picture quality and interference levels in tuner sections of TV receivers. Analog bandwidth of high definition video signals varies according to the scanning structure used, but generally is about two to three times the bandwidth of standard definition video.

In the true end-end-end digital TV world, analog channel characteristics, primarily bandwidth, need only be concerned with camera front ends and display inputs. In the latter regard, it might even be safe to say that one only need be worried about cathode ray tube displays because flat panel, non-CRT displays don't use analog driver circuitry.

Figure 1-3 is a block diagram of a representative video source showing basic functions and components, including optics, sensors, color correction, detail enhancement, gamma correction, and encoding.

Although this model is based on a live camera, film or telecine cameras used to capture images from film have a very similar architecture, except the image is transferred from the film to the sensors by passing a light source through the film to project image onto the light-sensitive surface of the sensors.

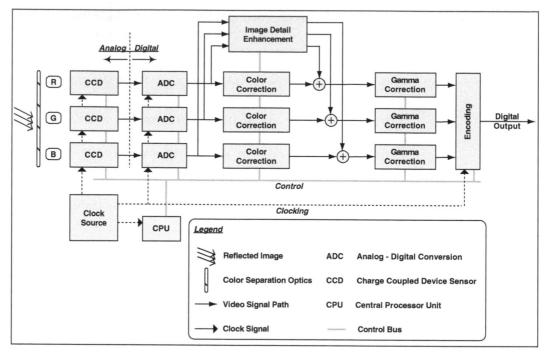

FIGURE 1-3 Video Source Block Diagram

Digital video is a product of a combination of the scanning and pix-elization structure used in most cameras and pickup devices used in broadcasting, security, and conferencing applications. Most modern camera designs use 27MHz clocking found in MPEG compression. It's beyond the scope of this book, but worth keeping in mind the fact that SMPTE and ITU 4:2:2 encoded, uncompressed video signals are parents of MPEG-2, 4:2:2 MP/ML compression family.

Planning and designing content transport networks requires detailed information about the payload bitrate and interface. For example, SMPTE 259 is a common interface for SDTV. SMPTE 292 is a common interface for HDTV. Equipment manufacturers design these standard interfaces into their products.

PACKAGING THE CONTENT SOURCES

Regardless of the nature of the various signals, sooner or later the signal will be compressed before storing, transported through a network, recorded, or sent out over the air. MPEG encoders are equipped with analog and/or digital input interface and output a compressed signal in a serial bit stream that is a fraction of the original bandwidth. It is this signal that is most often handed off to a satellite or fiber network. If it is live or previously recorded, it can be streamed or run in real time across the network. The recorded or stored version can be transferred as one would any ordinary computer file using file transfer protocol (FTP).

Program content is usually the product of multiple camera and microphone output signals mixed and switched to make continuous program material. In addition to these output signals, other material from videotape recorder/players—file servers, graphics systems, and remote, off-site feeds—can be used to contribute to any particular program segment. Figure 1-4 is a block diagram of a typical production system.

If there is a possibility of converging program content with voice and data traffic in a common network, most of the action has to be at

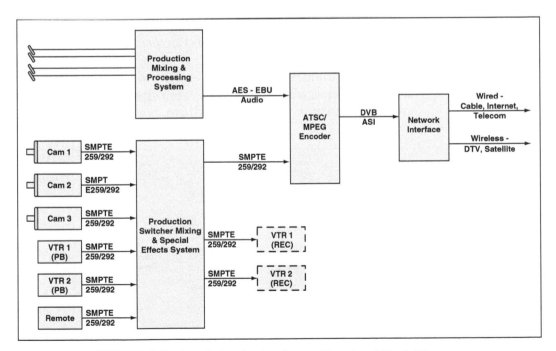

FIGURE 1-4 Simple Production System Functional Block Diagram

physical and virtual channel interfaces, and within a common understanding of bitrate and interface details. This is where good verbal and written communication skills and practices can achieve success, or a point where lack of understanding or misunderstanding leads to only partial success or outright failure. The idea is to define program content transport, in this case, audio and video signals, with coding, interface, and payload bitrates across common interfaces with consistent, standardized parameters. These details include the bandwidth required by the content payload, formatting and framing; coding, compression standards, interface standards, the payload bitrate and how it makes up a multiplexed signal in a package that can be transported across a common network access, transport, and egress facility.

Overall, the content creation, distribution, and delivery process involves at least one or more repetitions of several functions or processes in the sequence shown in Figure 1-5.

They include transformation, commonly referred to as *compression, source coding, multiplexing, channel coding,* and *media interface.* On the receiving side of the transmission channel, the compression, source coding, packetizing, and multiplexing functions are reversed and are called *demultiplex, parse, decode,* and *decompress,* respectively.

Coding techniques and standards are rules and practices detailing how bits, bytes, files, and streams are organized and mapped to and from the transport or storage medium. Source coding means that the

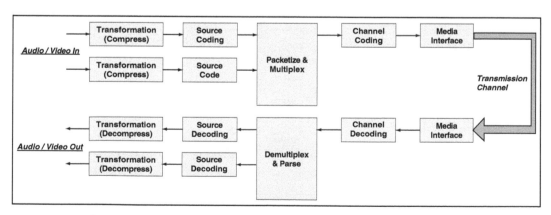

FIGURE 1-5 Program Content Transport Process

output from an A/D converter is in bits and the bits are arranged in a form and fashion whereby the information they represent can be decoded and made useful for all practical purposes as though it had never been encoded and decoded. Audio and video signals that have been encoded may be formatted into fixed or variable length frames.

Source coding and compression techniques may be lossy or lossless. Lossy means that information in the original signal is thrown away, or not included in the coding scheme. Lossless means the decoded signal is a bit-accurate representation of the original digital information, no more, no less. For example, the term 4:4:4 defines a complete set of digital bits, bytes, and frames from the A/D conversion process applied to the output of CCD sensors in a camera. The typical camera output is almost always stated to be 4:2:2, meaning the signal contains half as many chroma pixels as a 4:4:4 signal. Sometimes this technique is referred to as *sub-sampling*. Overall, it is part of a signal processing strategy called *bitrate reduction* (BRR) that reduces the amount of bits in the signal, lowering complexity and cost of camera processing circuitry and the amount of storage required of video tape machines and file servers, as well as transmission channel bandwidth.

Compression techniques and standards include algorithms, rules, and procedures used to process content before and after storage, transport, or both. The purpose of using compression is to transform the original information into something representative of the original that is satisfactory to the viewer or listener. Highly effective compression techniques strive to reduce bandwidth or storage capacity while preserving quality or resemblance to the original.

Interface standards define physical and electrical properties of connectors, voltage levels, signaling protocols, bit framing, and mapping details.

Payload bitrate is the data that travels across the interface. This includes all the data or bits resulting from analog to digital conversion, and any clocking, framing, control signals, metadata, or other directly related information. It excludes any and all bits or bytes related to network or storage media functions. To a communications network, or storage medium, 27MHz clocking information embedded in an MPEG compressed signal is payload.

Multiplexing techniques and standards are used when digitized audio, video or other source material is mixed or interleaved into a program level serial bitstream. Multiple programs multiplexed into a common bitstream are said to occupy virtual channels of the common channel or bitstream. Communications networks use still other multiplexing techniques and standards, some of which are statistical in their operational function. Generally, these techniques will not support content transport, unless it is properly packaged. For example, content packaged into Ethernet or IP packets and carried live or real time, requires a network capable of transporting and delivering the packets in the proper sequence and timing, and low levels of packet loss to be successful and satisfactory.

The payload bitrate across the interface between the broadcast operations facility and the network transport facility will probably be a serial interface. The capacity across the interface depends on several factors. Establishing the maximum capacity required is a simple matter of accounting for and summing up the basic parameters of each program and any required overhead included in multiplexing and other functions, such as error protection and correction.

Table 1-1 summarizes the range of payloads encountered when carrying high definition, standard definition, and streaming video typically found in creation, distribution, and delivery of program content.

ISO MPEG standards cover audio, video, and data signals compressed and multiplexed into a common output stream. The process adds overhead bits, increasing the size of the payload handed off to the network. It is also possible to combine or multiplex two or more

TABLE 1-1

Range of Payload Bitrates Across the Program Content Food Chain

Type of Material	Content Creation (Before Compression)	Content Distribution (After Compression)	Content Delivery (Before Decompression)
HDTV	1.2–1.5 Gbs	20–150 Mbs	12–20 Mbs
SDTV	270 Mbs	8–30 Mbs	4–10 Mbs
Streaming Video	270 Mbs Or Analog		300 Kbs–6 Mbs

Note: See Table 1-2 through 1-4 for more details regarding analog bandwidth, interface, sampling, and payload bitrates.

MPEG program streams into a common transport channel such as a cable, DBS, DTV, or telecom network. When program content is multiplexed into a single channel, this adds overhead bits.

Another source of overhead is forward error correction (FEC). FEC is typically added to the signal after compression but before mapping to the network transport channel. It improves end-to-end performance in terms of fewer errors in the program material caused by the network.

The bit rate of the audio or video signal after compression, stated in megabits per second (Mbs), is representative of, and directly impacts the payload bitrate. From the perspective of the network, content clocking information is part of the payload it must transport. The integrity and usefulness of the clock signal is dependent on the network. It must be kept stable within the limits of the decoder's ability to lock to it and perform its decoding function. Any jitter or errors induced into the payload stream, depending on their magnitude, can impair picture quality, sound fidelity, and validity of closed captioning or metadata included in the payload.

Another consideration is use of the content after it has passed through a network. For example, it's one thing to transport content through the network and use it for simple viewing on a TV set or a computer. It's an entirely different matter to use the same content as a feed in a production operation where it will be mixed with, or inserted into another program source. Two SDTV signals making up frames of 720 × 480 pixels, clocked at 270 Mbs must be synchronized and phased to match to within 740 picoseconds on a pixel-for-pixel, frame-for-frame basis, otherwise image impairment or outright loss of picture occurs. Similar ground rules apply to digital audio. For example, the ordinary CD source runs playback clocking at 44.1KkHz. Most professional audio systems and consoles run at 48KkHz. Using material from a CD without digital analog conversion requires re-clock and synchronization of the digital signal from the CD. Another issue, or potential concern, is synchronization of audio and video, sometimes called *lip sync*. This is nothing more than maintaining an accurate timing relationship between audio and video.

What is important from a communications network perspective is the bit rate and signal interface. The output of the audio console or mixer will get connected to a wire or fiber conductor or some type of network interface. A good example is AES3 serial digital interface used in television audio. The physical connector is a BNC coaxial connector and the encoded bit rate ranges from 192 to 756 Kbs for dual channel audio, scaling linearly upward for multi-channel sound payloads.

The second of our three basic models is the network interface device (NID). The basic function of the NID is as its name implies, interface the broadcast facility with the communications network. Figure 1-6 is a functional block diagram of a generic NID.

In simple terms, the NID enables content to be moved from the television world to the network transport world and back to the television environment, safely and securely, one bit, byte, cell, or packet at a time. Couple a well-designed NID with proper communications facilities and you have capability to interface and manage all the supporting voice and data communications, as well as managing end-to-end service for a simple point-point connection supporting content creation, or a multi-point network distributing and delivering content.

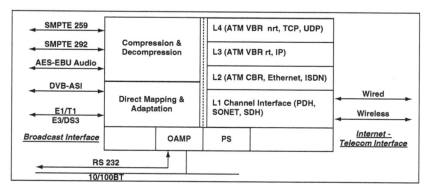

FIGURE 1-6 Functional Block Diagram Network Interface Device

The last model to be used is shown in Figure 1-7.

The purpose of this model is to provide the broadcaster with a tool to specify, quantify, and contract for equipment and services. With invocation of appropriate standards, competitive procurement becomes significantly easier. Once the goods and services are delivered, it becomes prudent to determine whether or not the deliverables meet

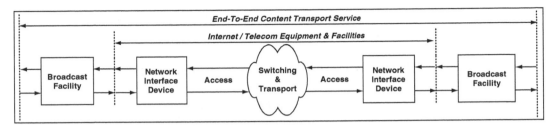

FIGURE 1-7   Inter-Facility Link End-to-End Service Model

the requirements of the contract, not just at the point in time where the delivery occurs, but continue to meet the requirements minute to minute of each hour, day, month, and year as agreed to in the contract.

In many situations, it's not possible to acquire network services from a single supplier. A well thought-out and documented end-to-end service model can be used to detail requirements for a **meet point,** a term for literally connecting one service provider's network interface to another. The end-to-end service model can eliminate finger pointing, and enable objective measurement and monitoring of individual components and elements of a network, not just the end-to-end connection. A well-documented, end-to-end design enables the broadcaster to separate facts from conjecture, and the known from the unknown during delivery and acceptance, as well as trouble reporting and resolution during operations.

ACRONYMS AND KEY TERMS

While there's a more extensive glossary of acronyms and terms in Appendix I, it is by all measures not comprehensive nor is it the only source you should have handy. Cisco has produced a very good dictionary[2] containing Internet working terms and acronyms. Gilbert Held, a well-known communications author has written an excellent data dictionary.[3] Another good source of Telecom and related terms is Harry Newton's Telecom Dictionary.[4] And there's always your favorite search engine on the Internet. Having said all that though, the last part of this chapter is an attempt to highlight the key terms and phrases you should review and understand as you read through the rest of the book and more importantly, develop your own set of knowledge and way of dealing with others where the subject matter is content transport in communications network environments.

First, a little philosophy, and background perspective:

Analog or Digital—Where did we come from, what are our individual backgrounds, and where are we headed?

If you're a technician or engineer it's likely your background includes some level of formal education in basic electronics. After school you begin practicing your new knowledge by installing or repairing equipment. Depending on your age, you learned about tubes, transistors, integrated circuits, large scale integrated circuits (LSIs), glue logic, central processing units (CPU), random access memory (RAM), read-only memory (ROM), and power supplies. Along the way, you came to appreciate not only the components themselves as individual pieces, capable of performing specific functions and by definition, incapable of performing lots of other functions better left to another component created to perform those functions or some limited set of all the functions in a larger system.

You may have started the learning process when the electronic world was mostly analog, or if you're just out of school and beginning to practice you're at the opposite extreme because the electronics world is mostly digital, or at least it seems that way.

There's another variation on this picture that may be helpful in understanding where we are and how we got to this point. It has to do with control systems. Somewhere along the way, electrical power found its way into not only powering mechanical systems, but controlling them as well. Control systems evolved along the same path as harnessed electricity did. *Batteries* produce only one polarity—direct current (DC)—and can be connected so the positive or negative terminals use a common conductor, or buss to provide power to the components. And the lowly switch, manually operated at first, then later automated with the invention of the relay, became the basic parts of control systems.

As alternating current (AC) evolved, control systems based on AC evolved. Ways were found to make relays open and close their switches with AC and perform the switching function just like DC-powered relays.

Control systems behaved much like today's systems simply because the power to control machinery was either on or off. And perhaps that's where the idea or notion of a *pulse* came into the picture, simply the application of power for a specific time. Control systems still exist today, but they are far more complex and capable of exceedingly complex operations. An example being the glue logic mentioned earlier, or something like simple network management protocol (SNMP), which is far from simple.

To survive, grow, and prosper in today's digital world you don't need to know much more about analog electronics. Analog circuits and systems are highly specialized and in limited use. They are likely to stay that way in the foreseeable future. If you're interested in learning more about analog audio and video, Luther[5,6] and Robin[7] are excellent references. Both authors have electrical engineering educations and television practice backgrounds. Luther's book gives a higher-level treatment of the subject, while Robin's is far more detailed.

Digital is the here and now and likely to stay that way in the foreseeable future, and there are some basics you need to know to survive. These include analog to digital conversion and digital to analog conversion, clock signals, pulse trains or digital streams, buffering and memory, parallel to serial conversion and serial to parallel conversion, bit stuffing and de-stuffing, timing and counting, system and network clocking, and synchronization. The Luther and Robin references also do a good job of introducing and teaching what you need to know about digital audio and video to deal with digitized content in network environments. If you are knowledgeable about digitized content as outlined in these two references, then you are well primed for transporting content in digital networks.

As a bridge to digital networks and digital content transport here's a list of topics and a little about each you need to know:

Analog to digital conversion (A/D) is the process of sampling a continuous wave voltage or current signal, and quantifying each sample to some precise real number mathematical value within a specific range. There are two steps in A/D conversion—sampling and quantization. A/D converters require power and clocking signals. The output of an A/D converter is always a group of signals, each on a single conductor, synchronized to the clocking signal. The number of signals appearing at the output of the A/D is always equal to the quantizing depth or capability of the converter. For example if the converter is said to be an 8-bit converter, its output signal would require at least 10 conductors; one for common or ground, one for clock signal and 8 for each of the signal bits ranging from 1 through 8 or 0 through 7, depending on the preference or convention of the designer. A 12-bit converter would require 14 conductors; a 16-bit converter would require 18 conductors and so on. The output of an A/D converter is called a parallel digital signal because all the bits appear in parallel timed according to the input clock signal.

Digital to analog conversion (D/A) is the reverse of A/D. A parallel digital input signal is converted to an analog output signal. Like its counterpart, the D/A converter requires power and clocking signals to perform its function. *Clock signal* is a continuous stream of positive and negative going voltage waveform transitions of equal periodicity. Clock signals are derived from highly stable continuous wave oscillators whose natural output is a sine wave. Typically the sine wave is of sufficient amplitude that it can be used to reliably cause the clock signal transitions to be precisely equal to and repeatable within a small range of time equal to less than 1 degree out of 360 degrees that define the sine wave. Classical television uses sync pulse generators to perform the same function.

Pulse trains or *digital streams* are a series of pulse edge transitions capable of conveying intelligence and information. Pulse trains, bits, or *bit streams* as they are often called, are organized into units called *bytes,* each byte consisting of 8 bits. Computers and communications systems deal with bytes as a common unit.

Buffering and memory means to use storage as a holding point for a collection of bytes of information and intelligence for any number of reasons, depending on the operation that's being performed. Operations using buffering and memory cause latency or delay in transport through a piece of equipment or a network.

Parallel to serial conversion means to take the parallel signal from an A/D converter and organize it into a series of bits so the information and intelligence can be transported on a single conductor sequentially, or in serial fashion. The serial bit stream contains clocking information and data, sometimes called *Payload*. The clocking information always appears in the same form and place to enable and maintain synchronization of clock and payload bits.

Serial to parallel conversion means the reverse of parallel to serial conversion. In other words, a series of bits in sequential fashion containing clock and data signals are separated and stored in memory or buffered for as long as it takes to separate and organize each of the payload bits or signals in the same time frame as the original clock and data bits and then place each of them on the output conductors in the same time frame with its original clock.

Bit stuffing and de-stuffing is the act of adding or removing bits in a network transport stream as a temporary measure to prevent or mitigate network clock slips.

Encapsulation means heading or surrounding a payload of bits or bytes with other bits and bytes for some functional purpose such as error correction, routing, or switching.

System and network clocking and synchronization can be a messy subject to deal with in theory, but in practice it's simple, especially in content transport networks. Basically, there are three clocks to be concerned about—the clock in the content origination facility, the network clock and the clock on the other side of the network receiving the content. Each is independent of the others. These clocks might be in phase and running on the same frequency within very tight tolerances, or they might be running at their design limit permissible extremes. If the network is transporting content in non-real time, or as a file or object, there's no need to worry about synchronization or timing relationships. The file gets moved according to ftp rules, a chunk of bytes at a time. On the other hand if the content is being transported in real time, the process involves moving the payload from the originating source into the network timed domain without error, transporting it without error, and then moving it from the network domain to the second program content domain without error.

Digital video broadcasting—asynchronous serial interface is an increasingly popular interface selected by designers for use in digital television facilities. Broadcast engineers with a good understanding of SMPTE-259 like it because it's a lot like what they already know. Communications engineers like it because it's a lot like E3, DS3, and STM1.

Facility means generally, something physical such as a building with equipment and amenities used to produce content or provide service. Examples include editing facility, production facility, studio facility, and transmission facility in the broadcast world. Telecom facilities are often called services. For example, it's quite common, and a misuse of the term to say or write *T1 service* or *DS3 service* when describing or meaning a T1 or DS3 access or transport facility. Similar to broadcast operations, communications operations take place in physical facilities—the building that houses operational functions. Then there's access facility, cross-connect facility, switching facility, terminal facility, hosting facility. Use of the term continues loosely, but has precise meaning in the community when it's used in the context of a private line facility. A more specific private line facility might have bandwidth characterization added such as seen in the term E1, J1, T1, E3, J3, T3, or the DS designation in front of the word facility.

Forward error correction refers to bits or bytes of information added to content that can be used to detect and reverse errors caused in the transport process. For example, terrestrial and satellite wireless (radio) transmission is said to rely on 3/4 or 7/8 FEC, meaning that for every 3 units of payload, add 1 unit or for every 7 units, add 1 unit of specifically coded information to be used to detect and correct errors in transmission.

Interface means a connector, or connection point, sometimes also called *service handoff* or *service demarcation*. Such a designation is most useful when it carries a reference to a single standard such as G.703, RS-232, DVB-ASI, or IEEE 802.11.

Units of Measurement:

Hertz (Hz)	1 cps (cycle per second)
Kilohertz (kHz)	1000 cps
Megahertz (MHz)	1,000,000 cps
Gigahertz (GHz)	1,000,000,000 cps

Channel Capacity or Payload Size:

Kilobits (Kb)	1000 Bits
Megabits (Mb) –	1,000,000 Bits
Gigabits (Gb) –	1,000,000,000 Bits

Cell, File, or Packet Capacity:

Byte	8 Bits
Octet	8 Bits
Kilobyte (KB)	1024 Bytes
Megabyte (MB)	1,048,576 Bytes
Gigabyte (GB)	1,073,741,824 Bytes

1 Kilobit is usually understood to mean 1000 bits, 1 Mb is 1,000,000 bits, and so on. But be careful with the *bytes* term when measuring or calculating bits in a file or time to move the bits contained in a file. File sizes are always in octal numbers. A 1-KB file actually contains 1024 bytes, a 1MB file contains 1,048,576 bytes. Other potential points of confusion abound with fixed size ATM cells, variable size Ethernet, and IP packets. ATM cells are always 53 bytes long, and more specifically in ITU documentation and practice each cell is usually referred to as an *octet* containing 8 bits, no more, no less. But Ethernet packets range in size from 46 to 1500 bytes. Internet packets can be between 68 and 65,535 bytes long.

Rate of Delivery or Movement:

> Kilobits-per-second (Kbs)
> Megabits-per-second (Mbs)
> Gigabits-per-second (Gbs)

These terms also used to refer to channel or link speed. Be careful because in content transport network design and operation, the two are almost never equal because of additional overhead bits required by error correction, or even still more bits used to map ATM, IP or TDM PDH into SONET/SDH channels.

KEY STANDARDS AND DESIGN PARAMETERS

Please see Table 1-2, Content Creation, Table 1-3, Content Distribution, and Table 1-4, Content Delivery and Other Applications, on pages 33 through 35.

TABLE 1-2 Content Creation

Application	Analog B/W	Sample Rate	Bits / Sample	Coding	Channels	Interface	Payload Bitrate
Audio							
Studio Mastering	20–20,000 Hz	48 KHz	16–20	AES 10–1991	56	BNC	125 Mbs
Digital VTR	20–20,000 Hz	48 KHz	16–20	AES 2/3–1985	2/4	BNC	384–768 KBs
Live Broadcast	20–20,000 Hz	48 KHz	16–20	AES 2/3–1985	2/4	BNC	192–768 Kbs
	20–20,000 Hz	48 KHz	16–20	MPEG2	2/4	BNC	192–768 Kbs
				Sub-Sample	Frame Rate		
Video							
Digitized NTSC	~6 Mhz	13.5 Mhz	8–10	4:2:2 YUV	29.94	SMPTE 259	143 Mbs
Digitized PAL	~8 Mhz	13.5 Mhs	8–10	4:2:2 YUV	25	SMPTE 259	177 Mbs
SD 720×480/576	~15 Mhz	27 Mhz	8–10	4:2:2 YUV	29.94	SMPTE 259	270 Mbs
HD1920×1080	~30 Mhz	74.25 Mhz	8–10	4:2:2 YCbCr	29.94	SMPTE 292	1,485 Mbs

TABLE 1-3 Content Distribution

Application	Analog B/W	Sample Rate	Bits / Sample	Coding Scheme	Channels	Interface	Payload Bitrate
Audio				Coding			
Digital VTR	20–20,000 Hz	48 KHz	16–20	AES 2/3–1985	2/4	BNC	384–768 KbS
Live Broadcast	20–20,000 Hz	48 KHz	16–20	AES 2/3–1985	2/4	BNC	192–384 Kbs
	20–20,000 Hz	48 KHz	16–20	MPEG2	2/4	BNC	192–384 Kbs
Video				Sub-Sample	Frame Rate		
Digitized NTSC	~6 Mhz	13.5 MHz	8–10	MPEG2 MP@ML	29.94	SMPTE 259	10–12 Mbs
Digitized PAL	~8 Mhz	13.5 MHz	8–10	MPEG2 MP@ML	25	SMPTE 259	10–12 Mbs
SD 720×480/576	~15 MHz	27 MHz	8–10	MPEG2 MP@ML	29.94	SMPTE 259	10–50 Mbs
HD1920×1080	~30 MHz	74.25 MHz	8–10	ATSC	29.94	SMPTE 292	30–60 Mbs

TABLE 1-4 Content Delivery and Other Applications

Application	Analog B/W	Sample Rate	Bits / Sample	Coding Scheme	Channels	Interface	Payload Bitrate
Audio & Speech							
CD	44.1		16	Coding		RCA	176 Kbs
Digital Audio Player	44.1		16	RDAT		RCA	176 Kbs
Dolby Surround					6	RCA	448 Kbs
Dolby Stereo					2	RCA	192
PC Music (Stereo)		22.1 KHz	16	WAV	2	RCA	88.4 KBS
PC Music (Mono)		10.5 KHz	16–20	AES 2/3–1985	1	RCA	10 Kbs
Telephone Speech		8 KHz	16–20	ITU G.703	1	RJ11	64 Kbs
Internet Speech		2 KHz		ITU G-723	1	RJ45	16 Kbs
Audio Conference							
Video					Frame Rate		
Digital STL				MPEG2MP@ML	29.94	BNC	10–34 Mbs
DVD				MPEG1/2	29.94	RCA	4–6 Mbs
PC Based Training				MPEG1	15–30	–	1–2 Mbs
Video Conferencing				ITU H.263	15–20	RJ45	384–768 Kbs

References

1. Telarc, International, Telarc Revisits the "Blast from the Past" with New DSD Technology, Newly Recorded Cannons, Carillons and Choruses http://www.telarc.com/gscripts/title.asp?gsku=0541
2. Author Unknown. *Dictionary of Internetworking Terms and Acronyms.* Indianapolis: Cisco Press, 2001.
3. Held, Gilbert. *Dictionary Of Communications Technology: Terms, Definitions and Abbreviations.* Second Edition, New York: John Wiley & Sons, 1995.
4. Newton, Harry. *Newton's Telecom Dictionary.* 18th Updated and Expanded Edition, Gilroy, CA: CMP Books, 2000.
5. Luther, Arch C. *Digital Audio and Video.* Boston: Artech House, 1997.
6. Luther, Arch C. *Video Camera Technology.* Boston: Artech House, 1998.
7. Robin, Michael and Poulin, Michel. *Digital Television Fundamentals: Design and Installation of Video and Audio Systems,* New York: McGraw-Hill, 1998.

2

INTERNET AND TELECOM: A BRIEF HISTORY

Alexander Graham Bell is credited with inventing the telephone sometime after Samuel B. Morse came up with the telegraph key and code, making smoke signals obsolete technology. Somewhere along the way, in the more recent past, computers learned to talk to one another over telephone lines. And then along came the Internet.

That's almost enough history for practitioners. However, a perspective on the past is well worth a quick read because understanding some of the history, especially since 1982, provides insight into the changes the Telecom industry has undergone and how that has, or will impact media and entertainment industry operations in the future.

Keeping our focus on practical considerations it makes sense to start somewhere in the early part of the past century. After all, radio and telephones came from similar inventive roots and had electrical or "electronics" in common. It's also instructive to observe that the two parted ways when digital electronics went solid state. With digital electronics—initially the switching transistor—devices could count and keep track of items or service transactions, calculate, or measure interesting, valuable operational and accounting characteristics of the business.

Someone figured out how to convert an analog signal into digital form, and the telephone world went for it with a vengeance. On the other hand, radio and television receivers didn't warm to digital techniques until well after integrated circuits with divide and multiply capability became cost effective for use in tuners. Other similar events along the way could be mentioned, but suffice it to say that broadcasting and consumer electronics lagged the telephone industry in adoption of digital technology by many years and did so, one painful step at a time.

The telephone industry was *digital* in nature from the start. For example, the basic business transaction unit, or service function, was and still is, a phone call. The line was in use or it was free. It could be used or one had to wait. One person could talk to one person at a time (until party lines were figured out). One operator could connect two parties to make conversation, later capitalized by the industry in the form of a telephone call.

Almon B. Strowger, a Kansas City undertaker, invented the stepping switch to allow customers to decide between his establishment and a competitor without undue influence of the wife of his competitor who just happened to be the local *telephone operator*. The stepping switch enabled the industry to continue in a mechanized and automated fashion, without operators, one telephone call at a time, the staple of long distance telephone bills.

Audio—the staple of radio—was analog and stayed that way. In 2003, radio broadcasters began upgrading transmission facilities to digital transmission.

Video might be characterized as digital in nature—at the very least it's highly structured. Pioneer inventors Philo Farnsworth and Vladimir Zworykin both conceived a fixed, repetitive scan structure. Farnsworth reportedly conceived his version while plowing successive rows in a field.

Historically, telephony, radio, and television share a common business life-cycle behavior. Still in relative infancy, the Internet seems to be evolving in similar fashion. The life-cycle includes three distinct phases starting with experimentation, moving into growth and consolidation, and finally into a mature state where the technology continues to evolve and the business segment makes a continuing contribution to the economy over a long time.

EXPERIMENTATION

Experimentation is at the root of technological success. Technological success forges reality from fantasy or the fantasy remains just that—an idea with little or no value other than personal satisfaction.

Experimentation with telephony and radio proceeded almost in parallel. Bell's invention of the telephone took place around 1875 to 1876 when he built models and demonstrated technical feasibility. His early working samples enabled people to talk to each other over wire conductors across distances that far exceeded the range of direct human speech and hearing. Experimental development would continue throughout the late 1800s and into the 20th century as patents were granted and the first services became available.

From the start, the telephone industry was dependent on lines, simply two pair of wires connecting two telephone instruments. Over time, it became obvious that if all the lines were connected to a single, centralized location where any telephone user could be connected to any other user on demand, service and usefulness would be greatly improved. With this capability, the central office (CO) came into being.

Ten to twelve years after Bell's initial experiments, Guglielmo Marconi read about and began experimenting with Hertz' work with electromagnetic waves. Marconi believed that magnetic waves could free telegraphy from the constraints of wire and cable. After significant development work in the 1880s, and a convincing demonstration of sending signals over water between the shore and an island in Bristol Channel 8.7 miles distant, he changed the name of his company, "Wireless Telegraph and Signal, Ltd." to "The Marconi

Wireless Telegraph Co. Ltd." He received an English patent on the wireless in 1896.

Farnsworth and Zworykin conceived *television* in the early 1900s. However, neither knew of the others work, and it was well into the 1930s before serious experimentation was started on today's so-called "analog" television system. Radio had become a commercial success both in terms of broadcasting and receiver manufacturing. In addition, the art and science of broadcasting audio or speech using electromagnetic waves was well known. Most of the experimentation was about getting pictures—"video" over the same medium.

Early telephone companies built and installed lines by stringing wires on poles or other convenient supporting structures where the right to use could be arranged. The poles or rights to use other supporting structure might be owned by the same company or by another entity. Over time, a collection of groups of private lines grew across the city and landscape. Ownership and rights to use these assets were traded, bought, sold, and bartered. Companies went into and got out of the business. Some succeeded while others failed.

Telephone lines were similar to the lines used by telegraph operators to transmit messages using a code conceived by Samuel B. Morse in the early 1800s.

Telegrams—written messages—were the end result of a process that started with a spoken or written message given to a telegraph operator who encoded it into dots and dashes and sent it to another telegraph operator using a telegraph key. The second operator received the dots and dashes, decoded it, wrote it out on paper, and delivered or had it delivered by a third party to the person or entity it was intended for.

Telegraphy was adopted by the railroads. Their rights-of-way easily and conveniently supported cross-country, intrastate, and interstate lines. Command and control of trains as well as switching trains to other tracks with speed and efficiency couldn't be accomplished without telegraphy. Can you imagine what it might be like to dispatch and control trains with messages delivered by the Pony Express? Over time, this capability migrated to public use in exchange for money and became a commercial business. In similar fashion, the telegraph key and typewriter were motorized and morphed into the Teletype machine and eliminated the need for the skilled *Morse Operator*. The Teletype machine spawned Telex (international) and TWX (Domestic) services that lasted well into the 1980s when they were largely replaced by facsimile technology standardized by the CCIT (later renamed ITU).

GROWTH AND CONSOLIDATION

Expansive growth followed by consolidation in the telephone industry occurred initially between around 1910 and some would contend, 1982. Telephone service in the United States was almost solely the domain of AT&T, or The American Telephone and Telegraph Company. AT&T was a large, highly regulated monopoly. In other countries telephone service was mostly a government utility, usually part of the post office, thus the acronym PTT for Postal, Telephone, and Telegraph service.

An example of successful consolidation and growth in the telephone industry involves the consolidation of the original Bell Telephone Company and New England Telephone Company into American Bell, Inc. On December 30, 1899, American Bell became a wholly owned subsidiary of American Telephone and Telegraph Corporation, a New York Long Distance Company, and a wholly owned subsidiary of American Telephone and Telegraph Company.

AT&T was incorporated in 1885 to manage and expand the long distance business of American Bell Telephone Company and its licensees. In 1899, it assumed the business and property of American Bell and became the parent company of the Bell System, yet another informal name.

The company grew and consolidated over many years as a legal regulated monopoly. It literally built a public telephone system in the United States with connections and relationships outside the United States that were the envy of the world for many years. The company was organized across three lines of business: local and long distance, (including international), telephone service and equipment manufacturing (Figure 2-1). Regulation of the company was centered on the prices it could charge for it's local and long distance service. The US Federal Communications Commission (FCC) dealt with inter-state and international pricing. State Public Utility Commissions (PUCs) regulated intra-state pricing.

Overall economics of the company turned on revenue produced by the local operating companies and the long distance business. The local operating companies built and maintained the local exchange networks. There were two basic parts to the local services business. Local telephone service typically included both the ability for one subscriber to call another in a common service area and service across a wider geographic area that was not deemed long distance. The second service or capability was also the point where long distance calls were handled, thus the name *access network*. Together, these two local service entities would later become the basis for creation of the *local access and transport area* (LATA) in a consent decree[1] with the US Department of Justice.

Because the local services business was regulated by the PUC in each state, these companies typically were organized and operated locally with unique names and local management. Examples include New Jersey Bell, Chesapeake and Potomac Telephone, Nevada Bell, Michigan Bell, and so on.

The long distance business operated under the name of "AT&T Long Lines." AT&T Long Lines, essentially a domestic US business, was

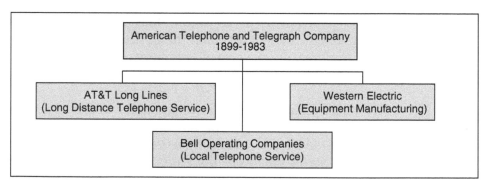

FIGURE 2-1 AT&T Business Entities Before 1982

responsible for building and maintaining the interstate long distance network and switching gateways to the international network. The long distance business was the first to deploy digital transmission and switching technology. The basic operation of the long distance network was simply to take a call from one local service network, connect it to a peer local company, and terminate the call. The originating company kept track of the date, time, and duration of the call and billed it as a separate item or "Long Distance Call." On a periodic basis, the entities separated the revenue, paid their suppliers, employees, and invested in new plant and equipment, all under the watchful eyes of the FCC and state PUCs.

Two other important, but less well-known entities were Western Electric Manufacturing and Bell Labs. Western Electric was the company's equipment manufacturing business. Essentially, this entity manufactured equipment and in later years developed software to run it. Its output was switching and transmission products and systems.

AT&T survived, grew, and prospered as the Bell System until 1984 when it was broken up by the consent decree. One corporate entity was converted overnight into eight. The long distance and equipment manufacturing business retained the AT&T brand; the seven new entities got the Bell name. The seven new entities Regional Bell Operating Companies—RBOC, for short—were given the rights to the famous Bell Logo, while AT&T got the rights to the word Bell as in Bell Labs.

On January 8, 1982, AT&T announced that it had agreed to break up the Bell System in response to Justice Department demands. This news was very unsettling to the company's 992,000 employees and welcome news to competitors, many of whom had been chipping away at its business for many years. Why? Different people will interpret the facts in various ways. However, most agree that the company's once unchallenged position had been seriously curtailed. Regulatory and judicial rulings had whittled away its ability to protect its business, starting with the Carterfone* decision in 1968.

At the time, the most lucrative part of AT&T's business was the long distance business. For many years, the operating companies or local exchange business had used their share of long distance revenues to make up for losses incurred in providing local service. Long distance competitors using Bell telephones as terminal points set up service at reduced pricing. AT&T long distance rates were set on average cost and included the subsidy paid to local exchange companies. Long distance competitors picked high traffic routes, leaving the lower traffic, higher cost routes to AT&T. Such action was referred to as *cream skimming* or *cherry picking*.

On another front, Western Electric was restricted from selling its output to any domestic customer except the Bell Operating Companies. Bell Labs, where the transistor was invented, had to make its patents available to one and all and was not allowed to use them in products other manufacturing businesses were making and selling. Competition was a one-way street. Other companies could use Bell facilities and discoveries to benefit their business, but AT&T was forbidden to start any new enterprise that might be construed as taking advantage of its size, skill, and knowledge.

Big and powerful though it was, the 22 Bell operating companies were not the only ones in the business. At the time, almost 1500 independent companies provided service to some 35 million subscribers. The 1982 Consent Decree did not apply to these companies, but they were impacted in many ways because of working relationships with various parts of the Bell System. 1983 was a year of planning and preparing for January 1, 1984. This was the date of start of business for the Bell Systems' 22 operating companies. Each entity was organized under one of the seven newly created *regional Bell operating companies* (RBOCs).

Carterfone and Carterfone Decision. Carterfone is a device that enables acoustic coupling between a telephone and a mobile or two-way radio. Strongly objected to by the Bell System, only a few thousand were ever made. Brought before the FCC, it provoked a landmark decision in 1968 when the Commission ruled that not only Carterfones, but also other acoustically coupled devices that were not owned or provided by the telephone company could be connected to the public telephone network if they were privately beneficial and not publicly harmful. The ruling led to establishment of Part 68 of the FCC's Rules. See Newton, Harry. *Newton's Telecom Dictionary*. New York: CMP Books, 2002.

Assets, liabilities, and equity of AT&T were partitioned off into eight new corporate entities. One share of stock was replaced with eight shares of the new entities. A single listing on the New York Stock Exchange turned into eight. AT&T stock now represented equipment manufacturing and long distance businesses. Seven new stock listings including Ameritech, Bell Atlantic, BellSouth, Nynex, Pacific Telesis, Southwestern Bell, and US West represented the local service business.

Two other actions worth noting involved the famous Bell Labs. In addition to splitting the Bell name and logo, the assets associated with Bell Labs were divided between the new AT&T and the seven regional operating companies. All research and development remained with AT&T's equipment manufacturing arm. A ninth entity was created and named Bellcore. Bellcore was essentially the part of the former Bell Labs responsible for standards development and important things, such as the North American numbering plan. Bellcore was set up as an independent entity with its own set of books; however, ownership was held by the RBOCs and managed by representatives of each of the RBOCs. This entity would later be sold off to a third party and renamed *Telcordia*.

The new AT&T consisted of administrative staff, AT&T Communications (Network Services), AT&T Technologies, (Western Electric and Bell Labs) and AT&T International (Figure 2-2).

The RBOCs lost all their share of long distance revenue except charges for intra-LATA calls not included in fixed monthly service charges, but retained the directory services business. They were permitted to sell equipment in competition with AT&T Information Systems, but restricted from designing and manufacturing equipment.

The seven RBOCs included Ameritech, Bell Atlantic, BellSouth, NYNEX, Pacific Telesis, Southwestern Bell, and US West.

Economically, the RBOCs were viewed as being in trouble. Charges for monthly service covering calls within a local access and transport area or LATA belonged entirely to the BOCs. The resulting revenue didn't cover expenses. Loss of so much of the long distance revenue made them unprofitable. Previously, state PUCs had refused to allow increases in pricing for local service. Now that long distance revenue

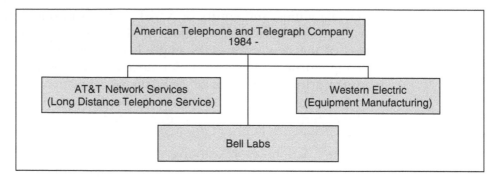

FIGURE 2-2   AT&T Post 1984

was gone, something had to be devised to make up for the loss in revenue. So the FCC created an *access charge* to be added to all long distance charges. The charge showed up on bills sent to long distance carriers and on subscriber monthly bills.

The RBOCs are permitted to sell new customer premises equipment that is not of AT&T Technologies sourcing. They cannot manufacture equipment but are permitted to enter other lines of business with restrictions.

Calls between LATAs or across LATA boundaries belong to the long distance carriers.

LONG DISTANCE DEREGULATION AND GROWTH AFTER 1984

The battle for the long distance market really started on August 13, 1969, when the FCC voted 4 to 3 in favor of granting MCI's application to construct a microwave radio system between Chicago and St. Louis. At the time of the vote, AT&T had a de facto monopoly on all long distance service in the United States. The vote to grant MCI a construction permit was the first signal of a breach in that monopoly. For the first time since the eve of World War I, AT&T had competition.

The breakup of the Bell System occurred on January 1, 1984, when it entered into a consent decree with the US Department of Justice. The terms of the consent decree prevented the RBOCs from entering the long distance or equipment manufacturing business. The decree also established equal access to the local exchange networks for AT&T and other common carriers such as MCI.

Before the breakup, state PUCs and the FCC controlled pricing for services based on return on invested capital. Prices for services were highly controlled and generally increased only when justified by a new or improved service resulting from a capital investment. The regulatory framework was based on intrastate services, the domain of the PUC and interstate services under the watchful eye of the FCC. Revenue funneled through the local exchange business just like the telephone calls. Intrastate revenue and interstate revenue funded the monopoly.

Part of the breakup was the establishment of more than 200 LATAs to implement the local access provisions of the consent decree. The local exchange companies could only carry traffic within the LATA, effectively maintaining their monopoly. Inter-exchange carriers carried traffic between the LATAs. Intrastate matters remained the domain of the state PUC, while interstate services continued under FCC jurisdiction.

From a practical standpoint, this new arrangement took time to understand and get used to. Each inter-exchange carrier established one or sometimes more points of presence (POP) in each LATA. These POPs continue to exist and provide access points in the LATA where they terminate and pick up traffic.

Ordering and provisioning service under the new way of doing business with the telephone company became something of a nightmare. If you were a consumer or small business requiring a single telephone line all of a sudden you were required to order local service from one source—the local exchange service provider who could not offer long distance service—and a long distance service provider who could not offer local service. No more one-stop service and, worse yet, two bills to pay at the end of the month where there used to be only one. Worse yet, those who had never experienced the joy of shopping for a telephone instrument could delay that for a while; however, avoiding paying exorbitant prices in a third monthly bill for equipment was impossible.

The other extreme represented by large businesses with multiple sites requiring service had the opposite kind of challenge. Before the break-up, many large businesses were served by a centralized group of people, typically located physically close to corporate headquarters and in many cases fully dedicated to serving the account. All of a sudden the former account management group had broken up into multiple groups, one for long distance and anywhere from one to seven or more for local service. Worse yet billing turned into a nightmare. Billing was site-specific. Corporations formerly paying a single bill for combined long distance and local service, covering hundreds and in some cases thousands of locations, had a blizzard of paper to deal with. (See the "Anatomy and Pathology of Communications Billing and Payment" section of Chapter 11 for more details on how the breakup impacted customers.)

As an example, billing for private line services used in host-to-host or terminal-to-host data communications resulted in three bills: two for the local access or tail circuits and a third for the inter-exchange facility. Ordering and provisioning private line services now turned into a coordination exercise for anyone requiring such service. The model for present-day service was set to the extent that local service is ordered and provisioned separate from long distance.

A local telephone call originates and terminates in the same LATA. A private line between two points in the same LATA is local service. The same local facility or service configured to terminate in the inter-exchange POP provides access to the inter-exchange carrier. The inter-exchange carrier provides service to the distant or foreign LATA. There it is terminated in a POP and passed to the local exchange carrier providing far end access, sometimes called *egress*.

The LATA scheme became the basis for widespread use of the terms *access* and *transport*. Functionally, the BOCs provide access facilities, and long distance carriers, also called inter-exchange carriers, provide transport.

LOCAL SERVICES DEREGULATION AND CHANGE AFTER 1996

With the breakup of the Bell System, long distance services were deregulated and began a path to a competitive marketplace. AT&T remained under greater restrictions than other common carriers. Regulations prohibited them from providing local service of any kind and they continued to be subject to tariff processes administered by state PUCs and the FCC.

The RBOCs were restricted as well. They were prohibited from offering any kind of inter-LATA service with three exceptions, the North Jersey LATA, an area on the border between Newark and New York; the South Jersey LATA between Philadelphia and Camden; and the DC LATA covering DC, and immediate surrounding areas in Maryland and Virginia.

With passage of the 1996 Telecom Legislation, the RBOCs were given a path to compete in the long distance market in return for opening up their networks to re-sale by third parties. This legislation defined and named entities in the local exchange business. The RBOCs were branded incumbent local exchange carrier (ILEC), and all others were named competitive local exchange carrier (CLEC).

THE INTERNET

It's hard to view the Internet as having been in an experimental stage. Technologically, what's now known as the Internet is a collection of myriad techniques from other communications methods and networks. Some might contend Internet experimentation started in the 1970s and continued into the 1980s. Initially the *Internet* was a US Department of Defense (DOD) research project aimed at developing survivable networks. Over time, participation in the research was expanded to include academic and private research institutions. By the mid-1990s, the DOD had its survivable network. Responsibility for various administrative and technical functions previously held by the military was turned over to non-government entities.

Historically, and by example, telegrams—maybe even smoke signals or pounding on drums to send and receive coded messages—could be considered "data" communications. Telex service remains in use in some parts of the world. It had a counterpart in the United States called TWX. Telex service is an international service, whereas TWX is or was restricted to domestic US locations. Both services are enabled with a *teletype*, a machine created by combining telegraph key and typewriter technology. Telegrams are a service offering used when a teletype machine is used to create, transmit, and print a message for a third party. Telegrams ceased to be used in the early 1990s.

Telegrams, Telex, and TWX effectively died out in the 1990s. AT&T is said to have transmitted its last telegram in 1991. So the cannibalization of the telegraph by the telephone took almost a century.

It's almost too complicated to go into, but along the way, AT&T and the RBOCs had, and continue to have, great challenges with anything outside telephone or *voice services*. The combined entities had deep and credible resources and capabilities without equal, except perhaps IBM. However, getting these resources to market was impossible because restrictions on AT&T equipment manufacture prohibited sale of their products to any other than the former BOCs. One result of divestiture and deregulation freed AT&T to sell computer equipment in competition with IBM and others. But coincidentally with the historic event in the telecom industry, Intel and Microsoft blindsided and blew away the dreams of the white-coat, MIS department mainframe computer industry, and those who would see AT&T as a possible competitor to IBM.

Setting all the issues of selling computer or computer-like equipment aside, the former AT&T and new entities had been and would continue to be dominant in a small and growing market for what started out as *computer communications* or *data services*. After all, AT&T long lines were first to install digital transmission and switching equipment in their network. Earlier FCC decisions had clearly and cleanly marked a dividing line between equipment and services. On the communications side a piece of equipment specifically defined as a *customer services unit* (CSU) faced another piece of equipment called a *data services unit* (DSU). The CSU was part of the computer system. The telephone company installed the DSU and included an amount in the monthly service charge. These functions are now built into equipment the customer buys and the telephone company installs at the customer's location or nearby equipment cabinet.

Data communications has become a matter of buying compatible equipment and services. In a few short years, the Internet has taken hold across business and consumer or residential services like nothing in the past. Along with that change has come tough times for investors and employees, and opportunities for users to avail themselves of new capabilities. Now, and for the foreseeable future it's "voice, data, and video."

CONSOLIDATION REDUX

From approximately 1999 through the end of 2002, telecom employees and executives saw option packages worth tens of millions of dollars, and their jobs, evaporate. Some of the most prestigious companies in the business were thrown into turmoil. Lucent Technologies, parent of Bell Labs, shrunk its global workforce from 153,000 to 35,000.

The nation's top telecom clusters in New Jersey, San Francisco, Boston, Dallas, Atlanta, and Washington suffered as well. High-flying startups initially flush with venture capital ran out of money and simply ceased operations or declared bankruptcy.

In early 2003, a deeply divided Federal Communications Commission considered and left in place rules passed in the wake of the 1996 Telecom Legislation that were intended to foster local telephone competition. Seven original RBOCs had consolidated into four. Ameritech and Pacific Telesis became children of SBC. Bell Atlantic merged with Nynex. GTE, once independent, that is, not part of the original AT&T, or a Regional Bell, merged with Bell Atlantic to form Verizon. Ostensibly, this action was to free the Bells so they could foster growth in broadband services. Freeing the Bells ideally would have allowed them to stop making their network elements available to resellers, allow them to compete in the long distance market as well as offer Internet services, the new name for the old "data" services. Part of the rationalization for allowing the Bells to compete in broadband services, allows them to compete with cable companies' Internet service offerings. However, this ideal was not to be.

The structure of the new rules removes restrictions on broadband, also known as high-speed Internet access, but not traditional telephone service, part victory and part setback because the Bells' traditional telephone service will remain subject to regulation of state PUCs, and thereby likely to continue to require the Bells to make their network elements available for resale by competitive carriers. So it seems like the world, at least in the United States, the communications landscape will continue to include the courts, various regulatory bodies, and to quote FCC Chairman Michael Powell, "Picasso-esque," predicting it will lead to "legal and regulatory chaos."

Where is all this turmoil leading? It's difficult as always to predict, but it's not difficult to get an indication from a simple look at the status of the local exchange business from the perspective of the dominant and not so dominant players.

First, the dominant players are generally considered to be the ILECs. The former RBOCs have for many years prospered from voice services sold to enterprise and residential customers. For the most part, these services are delivered via single-pair copper wire. Even where the services are delivered via fiber, the nature, character, and performance of the facilities delivering the service is bound by the constraints inherent in channelized, 64-Kbs T-carrier or PDH access, switching and transport hierarchy. Unchannelized facilities are available, but are simple point-to-point private lines. Digital interface to, and transport through, ILEC network facilities is limited to T1 and DS3.

A financial glance will show suffering from declining revenue and high debt. Declining revenue from wired voice services—their bread-and-butter—is being made up by mobile or wireless voice services. Overall, opportunities for growth are constrained geographically and technically. Relief from the geographic restraint provides opportunity in the form of incremental revenue from long distance services, but it is in exchange for unbundling network elements for sale to competitive local exchange carriers (CLECs), who turn right around and bundle and sell these same network elements in the same market, depriving them of pieces of their most profitable revenue streams.

Many of the ILECs profess to be interested in broadband equipment and facilities. Establishing significant levels of service requires significant investment in new equipment. There's been an initial "toe in the water" in the form of digital subscriber line (DSL) equipment acquisition and service offerings. But for anything significant such as would allow them to take market share from broadcasting, cable television, or satellite operators, the investment would have to be fiber replacement of copper wire, with IP access, switching and transport, in essence a wholesale replacement of existing infrastructure over time and that wouldn't necessarily relieve them of their regulatory requirement to continue selling unbundled network elements (UNE).

On the less than major player side of the market, there are the CLECs mentioned previously. CLEC business strategy comes in two forms: one sells Internet access and the other buys and sells unbundled network elements mentioned previously. Others, and as of the publication time of this book (early 2004), there are a few who have invested in their own facilities in large markets where the investment level is potentially rewarding because of short transmission distance and access to the more lucrative enterprise customer. One or two of the cable television operators have installed SONET/SDH equipment on common fiber with TV service and offer voice service in the residential market. They have been moderately successful, but real returns are only incremental mostly because of the limitations imposed by voice service pricing in a monopoly market dominated by incumbents.

Many are betting voice-over IP (VOIP) will be at the roots of the next major wave of change in the industry. Some maintain that this could be the next utopian storm. And it may come at faster speed than the classical telephone world is used to enjoying. Remember the game you learned as a child whereby a series of dots were placed on paper in a matrix form and then you and another player started connecting two dots at a time with a single mark? The objective of the game was to reach a point where the last line drawn created a box into which the player could put their initial and claim ownership. The end of the game came when all boxes had been claimed, and the winner was the player who had acquired the most boxes. In the VOIP game, there's a lot of dots out there now, including laptop personal computers (PCs) with a microphone and speaker, to say nothing of the number of desktops with circuitry and plugs, even microphones and speakers thrown away or languishing in a desk drawer. Microsoft started shipping operating system (OS) software with QOS capabilities in 2000. The next generation OS includes what they called a *real-time communications* client.

PC technology is pervasive or becoming that way in enterprise and consumer market places. Just connecting these dots alone could make a devastating impact on ILEC voice service business. Sooner or later, the owner of a PC will discover that many or most of the other people they talk to on the telephone can be reached by an alternate route through their PC and Internet access and wonder what else they could buy with the $30 or $40 they spend for local and long distance telephone service.

Local area network (LAN) technology and products have been as much a business necessity as telephone products and services for many years. Their utility started out as a way to share printers and later files. Somewhere along the path, the MIS department figured out they were a cheap replacement for the coaxial cable used to connect expensive terminals to mainframe computers. When the Internet came along, what did the business connect it to? For sure they didn't connect it to the telephone system.

Cable television operators discovered cable modems could provide Internet access. This immediately made a significantly positive impact on revenue from an incremental investment, far less than the incremental investment required to offer telephone service using SONET/SDH equipment. These devices connect to a PC, not a telephone.

BROADCASTING AND THE INTERNET

"Fortunately, nobody owns the Internet, there is no centralized control, and nobody can turn it off"—Saltzer et al[2]

Conceptually, the Internet is seen and promoted by its proponents as a wholesale replacement for classical telecom networks. Their concept of the Internet is essentially a dumb pipe or set of pipes, any of which can be used to provide end-to-end service without a single point of failure under control of, and at will of, the user. They see it as a utopian replacement for the telecom network, which they view as a rigid, monolithic entity made up of interfaces with the user on one side and control of services by the network provider or carrier on the other.

As long as it's here and established, we should relax and enjoy it, right? Well, yes and no. Yes, when it's practical and makes sense, and no, when it's impractical and doesn't make sense. To some, the Internet is anything and everything anyone could ever possibly want. Sure it's valuable and society benefits greatly from it. From a classical broadcasting perspective, it's a mixed bag, and like anything new on the overall landscape from smoke signals and drum pounding to modern communications, it represents threats and opportunities.

Perhaps an appropriate way to view the Internet is to see it as another way to move information and maintain contact between us human beings. The Internet is another new phenomenon on the continuum from smoke signals and drum beating through writing and drawing on cave walls and rocks to newspapers, books, the telegraph, movies, magazines, telephone, radio, television, and computers.

The Internet is simply another disruptive technology or capability. It will continue its disruption for the foreseeable future. From a broadcaster perspective, radio disrupted publishing and movie audiences. And television disrupted radio audiences worse than it did movie patronage. Television news had a different impact on newspaper-delivered news than it did on radio news, but it affected both. In the mid-1970s, two disrupters came along. One was the videocassette recorder (VCR) and the other was Home Box Office (HBO). VCRs changed network television viewing habits. All of a sudden, the major networks viewing audiences began declining and aren't likely to reverse trend anytime soon. HBO put movies on a satellite transmission facility and paid for receivers to be installed at cable system head ends in return for a share of the cable operators' revenue, from an extra charge, or increase in basic charges for service. Once the physical infrastructure was in place, *super stations*—Turner, WGN, and others—found it easy to uplink the program material broadcast over their local terrestrial transmission facility. All of a sudden, the local cable operator represented more viewing choices for the television audience, further fracturing the major networks audiences.

The Internet is a threat to the broadcaster's business. It's another choice for the listener or viewer. In office buildings where satellite reception is impossible, and radio and television reception is poor and borders on un-usable, the Internet provides a way to get music, news, and information to people while they work. A perfectly usable weather radar picture can be viewed with plenty of space on the screen for commercials. Handheld, battery-powered devices capable of pictures, sound and text creation, transmission, and reception represent threats to the telephone company unwilling to respond to change as much as they do the broadcaster. All of a sudden, the newspaper has an equal opportunity for the valuable eyes and ears of consumers of news and information.

In summary, the Internet is a new transmission path, a new threat, and another less expensive and more ubiquitous way of gathering or sourcing content.

SATELLITE SYSTEMS AND TECHNOLOGY

A background summary on communications wouldn't be complete without including satellite systems and technology, another unique segment of the field. The global satellite system today has evolved over more than 40 years. Satellite services are grouped into fixed satellite service (FSS) and broadcast satellite service (BSS) by the ITU. The more common informal reference for the BSS is *DBS*, meaning direct broadcast service. In addition to communications, satellite systems and technology provide vital weather information, mapping, location information through the global positioning system, plus many valuable services to the military.

The current DBS system is conceptually very similar to one Arthur C. Clarke described in an article[3] written in the fall of 1945 for *Wireless World*. In this article, he foresaw 24-hour manned satellites being used to distribute television programs. Despite a repeated version of the concept in another publication, *The Exploration of Space*[4] written in the early 1950s, the idea never gained much interest or attention.

John Pierce of AT&T Bell Labs is credited with being the first to take serious technical and financial interest in the idea. Pierce elaborated on the basic idea to the extent that the space-based platforms would perform much like a mirror and be located in medium and 24-hour orbits. He estimated the capacity of the satellite to be equivalent to 1000 simultaneous telephone calls and comparing it to the first trans-Atlantic telephone cable with a capacity of 36 simultaneous calls, arrived at a conclusion that it would cost 36 million dollars and be worth a billion.

AT&T caught the FCC by surprise in 1960 when it requested permission to launch an experimental satellite. At the time, the commission and other parts of the government simply weren't equipped with policy and rules covering satellite communications. RCA was awarded a contract to build a medium-orbit satellite in mid-1961. Around the same time, Hughes was awarded a contract to build a high orbit, 24-hour satellite. By 1964, four medium-orbit and two high-orbit satellites had operated successfully. The Communications Satellite Act of 1962 formed the basis for Communications Satellite Corporation with an initial capitalization of 200 million dollars to build a system of several dozen medium-orbit satellites. Ultimately, COMSAT decided to build satellites for the higher geosynchronous orbit, the first of which was launched from Cape Canaveral in April 1965.

A key early broadcast event was televising part of the 1964 Tokyo Olympics. At the same time the United States was gaining this initial expertise and capability, other countries had been involved from the beginning. American companies built COMSAT's initial satellites and launch vehicles. AT&T negotiated with Foreign PTT organizations to build earth stations and began tests and experiments aimed at providing telephone service. By the time COMSAT's first satellite was launched and ready for service, France, the United Kingdom, Germany, Italy, Brazil, and Japan had operational earth stations. In August 1964, agreements were signed to create the International Telecommunications Satellite Organization (Intelsat).

By 1969, when Apollo 11 landed on the moon, half a billion people watched the event all over the globe through Intelsat transmission facilities. The last facilities making up the first global network were placed in service over the Indian Ocean just days before the moon landing occurred on July 20, 1969.

ABC proposed a domestic satellite system to distribute television signals in 1965, but it never gained traction. In 1972, ANIK was placed in service by Telesat Canada to serve the vast regions of the country. RCA and Western Union both launched the first domestic satellites in 1974 and 1975. AT&T launched its first domestic satellite in 1976. Satellites were intended to provide voice and data service; however, television quickly became a major user. By the end of 1976, 120 transponders were in service, each capable of 1500 telephone conversations or one TV program. Movie channels and super stations were made available to cable head ends, driving the growth in cable TV demand. During this same period, the major radio and television networks began using satellites to distribute programming to their affiliates. Satellite distribution would prove far more reliable and less expensive than terrestrial networks.

Arthur Clark's vision of watching television from a satellite would be realized in the fall of 1994 when Hughes, RCA, and Hubbard Broadcasting launched the DirecTV transmission system. The first serious competition for cable got off the ground. A few years later, Echostar would launch its Dish Network.

A key component of satellite technology, the traveling-wave tube (TWT) was invented in England and perfected at Bell Labs. It is used to generate the signal transmitted from the ground to the satellite and back from the satellite-to-ground station receivers. Achieving adequate power level for the signal to be received by the satellite and re-transmitted back to earth required very large (100-foot diameter) dish antennas in the early uplink transmission systems. Early TWT power output levels were only approximately 1 W, but they have grown to more than 300 W. Uplink antennas approaching one tenth the size of early versions now cost around 30,000. Receiving antennas that are the size of a large pizza now enable reception of several hundred TV programs and data links to millions of businesses requiring credit card authorizations and accurate inventory tracking.

When COMSAT launched its first satellite in 1965, it provided almost 10 times the capacity of the submarine telephone cables for almost one tenth the price. Telephone service on a satellite facility suffers from the long path it must travel. In the early days, the availability of the service was its key selling point. Satellite telephone service is still the service to and between many countries today. The first fiber cable, TAT-8, was laid in the Atlantic Ocean in the mid-1980s and provided competition. Satellites are still competitive in many applications, especially point-to-multipoint service such as DBS and network distribution to affiliates.

References

1. *United States v. Western Elec. Co., 1956 Trade Cas. (CCH) § 71,134 (D.N.J. 1956).*
   Note: In 1956, AT&T and the U.S. Department of Justice entered into an agreement called the "Western Electric Consent Decree," which settled an antitrust action filed in 1949 against AT&T for alleged anti-competitive practices. In 1974, the U.S. Department of Justice (DOJ) filed an antitrust action against AT&T et al. for violation of federal antitrust laws. On August 24, 1982, the D.C. District Court (MFJ Court), under U.S. District Judge Harold Greene, accepted a proposed settlement of the cases styled as a "Modification of the 1956 Final Judgment" (MFJ). The MFJ Court vacated the 1956 Western Electric Consent Decree and replaced it with the MFJ. A plan of reorganization for AT&T was submitted to the MFJ Court as a condition of this settlement. The divested Regional Bell Operating Companies (RBOC) were bound by the provisions of the MFJ, which limited them to providing local exchange services and, as originally entered, expressly prohibited them from providing inter-exchange services, information services, or manufacture of telecommunications equipment. The MFJ or consent decree required them to provide equal access to their network by inter-exchange carriers and information service providers upon a *bona fide* request.
2. Carpenter, Brian. *Internet Architecture Board (IAB) Architectural Principles of the Internet.* Internet Engineering Task Force (IETF) Request For Comments (RFC), 1968. hftp://isi.edu/in-notes/rfc1958.txt
3. http://www.lsi.usp.br/rbianchi/clarke/ACC.ETRelaysFull.html
   Note: This is only one of many sites resulting from an Internet search: "Arthur C. Clark"+"Wireless World 1945." Typically, the result is a scanned copy of the original article from the publication
4. Clarke, Arthur C. *Exploration of Space.* New York: Pocket Books, 1979.

3

CONTENT TRANSPORT NETWORK DEFINED

Before going into the details of content transport networks, it is important to not lose sight of what we are trying to accomplish in terms of the end-to-end service required to support the program content food chain cycle. Figure 3-1 is representative of a framework for the current and foreseeable future.

This framework is intended to apply to digital radio and other forms of content carried by the Internet and listened to or displayed on desktop machines and handheld devices with wired or wireless connectivity. The notion behind this sketch is to add details about how the content moves within and between production and transmission facilities and is finally delivered to the user or viewer.

This chapter is a high-level description of functions, capabilities, and limitations of networks capable of carrying program content. It includes a high-level architecture concept and shows how standardized communications equipment, facilities, and services can be used to fashion content transport networks.

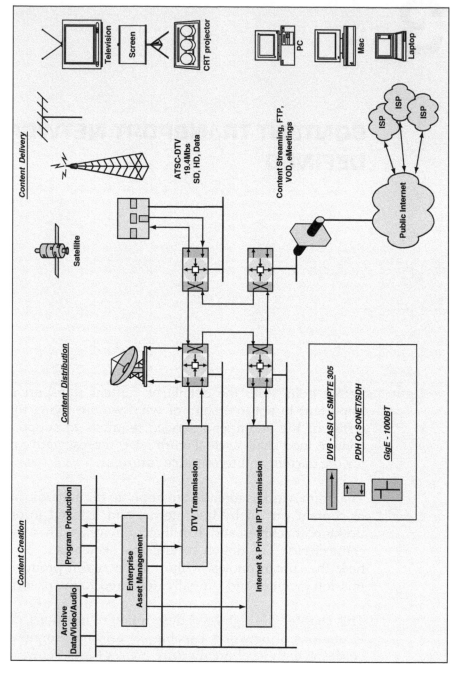

FIGURE 3-1

Say *ATM*, and most people think money, except those who know the meaning of asynchronous transfer mode. To carry this seemingly meaningless reference further, stop and think how often and in what context *network* is used. A network of friends getting together for snacks and drinks is an entirely different network than a food distribution network, or a communications network. It's not difficult to find many terms and phrases using the word. For example, voice network, data network, global network, ATM network, packet network, satellite network, radio network, fiber network, global network, and international network to name a few, or may be more apt, cable network, broadcast network, or the ABC, CBS, NBC, PBS, HGTV, CNN, and FOX networks.

Once again, we're back to the point of defining terms and then using them deliberately, consistently, and with purpose when discussing or documenting concepts and details of content transport networks.

Network-based content transport requires the network to be designed to meet the objectives of the business or enterprise owning or acquiring the content. The primary interests of owner and acquirer include availability, reliability, robustness, and security. A network designed to provide these basic capabilities can be called a *content transport network*. Put economics into the mix and it enables the designer to devise and offer choices or alternatives to the customer or user based on capabilities across a price range.

Ordinary, run-of-the-mill, voice and data networks generally don't do a satisfactory job of transporting content for a variety of reasons. However, voice and data services for a properly designed content transport network become an incremental piece of cake.

Properly designed content transport networks are, or should be, covered by a contract between a financially viable buyer and a technically competent, competitive service provider. To be effective, the contract must spell out requirements in terms of functional deliverables, including specific service characteristics, sometimes referred to as a *service level agreement* (SLA).

Both parties to such an agreement will be well served by a two-phase process. The first phase includes defining and provisioning facilities and services, followed by a test and acceptance process to certify the deliverables meet or exceed performance as outlined in the contract. The second phase is commencing and continuing operations with monitoring procedures based on strategically located monitor points.

Cost-effective, mission critical content transport networks become even more effective when relying on judiciously applied standards. Many standards bodies, including AES, ATSC, DVB, EBU, ETSI, IEEE, ITU, ISO, SMPTE, T1, and perhaps others, develop and maintain standards that can be leveraged to advantage.

In general, content transport networks exhibit high levels of availability, reliability, and robustness, and low levels of bit errors or high bit-to-error ratio (BER) performance.

Availability means that the network resource is available for use as specified in the terms and conditions of the contract.

Reliability means that the network will operate within the limits of the performance and headroom characteristics specified in the contract.

Robustness means that when an outage or out-of-performance condition occurs, the network detects the condition and restores the level of service within the timeframes specified in the contract.

BER is a measure of the bits transported without error, compared to the number of transmission errors over some specific time.

THE NATURE OF TELECOM NETWORKS

Telecom or common carrier networks have evolved over many years. The basic elements include switching, transmission, and network management. Telecom network operations run 24/7 and require very similar levels of flexibility and reliability as broadcast operations. Telecom networks are said to be ubiquitous and are built to reach the widest possible market for their services.

A key characteristic of communications networks is the concept of channelization or use of one or a group of channels between two points to support movement of information. Channelization first appeared when Edison and a now nameless technician accidentally discovered movement of sound waves over a pair of wires. That first pair of wires eventually turned into two pair to facilitate a two-way talk path. Over time through the magic of technological evolution ways and means to enable multiple channels on a single talk path, or four-wire facility were realized. For many years the jellybean of telephone technology was and still is the *voice grade channel*. The channel, be it wire or a virtual channel has certain capabilities and limitations bound by the laws of physics. Attempting to send a 30-Mbs payload through a 64-Kbs channel doesn't result in any more success than an attempt to pump 30 barrels of oil per hour through a 1-inch pipe. Expecting networks to carry voice, data, and video without some way to match each with a unique part of a communications channel is the equivalent of mixing oil, water, and orange juice into the same pipe and expecting each to arrive intact at the other end. The challenge is not so much in the mixing as in the separation at the receive end.

Telecom network architecture is standards based. Most telecom network operators standardize and base their designs on a limited number of manufacturers and suppliers, generally constrained by maintaining sound, competitive procurement practices. Equipment and software suppliers active in the market generally participate in standards development and tend to comply in most, if not all, aspects of applicable standards.

Telecom networks interoperate across business entity and physical boundaries that are local, regional, national, and international.

Telecom networks are built using a layered architecture. This architecture is based on international standards and consists of a physical layer, facilities, and service layers. Although this layering characteristic is related, it should not be confused with the seven-layer ISO model commonly used in data communications and information technology documentation. It is also related and should not be confused with the four-layer approach sometimes seen or referred to in Internet or Internet protocol (IP) documentation.

One of the main objectives of this book is to bring clarity to this picture for the reader. Clarity comes from understanding a greater level of detail of the telecom entity. For purposes here, the Internet is another facilities or service layer in the overall architecture. The Internet relies on classical telecom physical or transmission layer infrastructure. It is likely to remain that way for the foreseeable future.

Many, if not most Internet people, *netizens* as they sometimes refer to themselves in a third-person way, lack appreciation and understanding of the physical layer or the entire classical telecom infrastructure. To many it's old, outdated, obsolete, and subject to complete disregard, yet it is critical to successful operation of the Internet. That's because the original Department of Defense Advanced Research Projects Agency project included an assumption that the physical layer would always be 100% available. That general approach and attitude remains today and is reflected in such things as contemporary certification training and testing by router equipment manufacturers. There has always been a similar attitude and approach in classical data communications network design and operations. The simple message in this point is "don't forget the layer 1 and 2."

Another area of misunderstanding and common misuse is the term *private network*. Networks are made up of individual elements of hardware and software, supported and managed by a sophisticated network and equipment management infrastructure capable of monitoring, detecting, and reporting network behavior and performance outside predetermined limits. Strictly speaking, a private network is built using the same or similar kinds of equipment, software, and supporting infrastructure. It requires capital investment and ongoing operations expense. In the same context, private networks are used exclusively by their owners and don't provide services or facilities to others. Satellite facilities are generally thought of as private networks; however, regardless of the term, they are built using shared and non-shared equipment.

Public networks, also called *common carrier networks*, are shared among a community of enterprise and residential customers. They operate under rules and regulations promulgated by the Federal Communications Commission (FCC) and state Public Utilities Commissions (PUCs), based on state and federal law in the United States, and similar government bodies in other countries. This includes physical aspects of orbital platforms, radio frequency spectrum, and, to a lesser degree, the equipment and services in ground station facilities.

Virtual private network (VPN) is a term used to describe a network designed and created from public network facilities and services. In some designs, the customer purchases or otherwise acquires the right to use equipment and software making up a VPN. Typically, this equipment is located on customer premises, although it may be co-located in service provider facilities with other customers or carrier owned equipment. VPNs didn't just show up yesterday. The incarnation in the mid-1980s was *software-defined network* (SDN). These terms are another source of confusion. At a high level, both mean the same thing. Both evolved from private networks, which are built using point-to-point private line facilities and premises based multiplexing and switching equipment. The use of the terms *virtual private* and *software-defined* came into vogue around the time of the divestiture of AT&T and the deregulation of the long distance industry. Early examples of SDNs include AT&T and the Bell Operating Companies, who shared central office switching as an alternative to a private branch exchange and enhanced private switched communications services.

VPN is often used to describe a data network, but it is nothing more than a combination of point-to-point private leased lines and shared core routers or, in the old days, multiplexers. Some will claim that VPNs include encryption, firewalls, and so-called tunneling and that they are capable of carrying voice traffic. These claims and attributes are all true; however, the concept of shared faculties is the same as used in the SDNs that arose in the late 1970s and early 1980s to support multiple locations private dial plans for telephone service.

UNDERSTANDING TELECOM INFRASTRUCTURE

Much like broadcast systems and facilities, telecom systems and facilities have their unique symbols, labels, and documentation conventions. The key to understanding the technology and its application is getting a grip on the language. The keys to designing, specifying, and getting what you want include a basic knowledge of the subject matter and some design tools. The design tools include a computer with word processor, spreadsheet, and drawing capability. For large complex systems and networks, a database manager becomes an important tool for thorough and accurate cost and operations analysis. In addition to these tools, design reference material in the form of recommended practices, standards, product specifications, and data sheets are a must.

Another way of defining telecom infrastructure is a technique called *layering*. Layering, as used in computer and communications, is best explained as a technique whereby software and digital networks are designed and built in layers. The notion is that if the operational routine or software system is built in layers and each layer interacts or interoperates with the one above and below it, then the entire system is more likely to achieve its overall goals and effectiveness.

Examples of layering include the ISO Open Systems Interconnect model, the four-layer Internet model, and the synchronous optical network/synchronous digital hierarchy (SONET/SDH) four-layer model. These models are three entirely separate models, and while they may not have been created without awareness of each other, taken literally, they don't appear to have any relationship. Figure 3-2 shows these three models side by side and how they relate to each other.

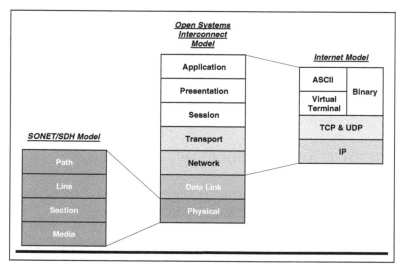

FIGURE 3-2   Open Systems Interconnect Stack

It's almost impossible to design, build, and operate a network without some form of integration of all three. These models will be applied in detail in the Chapters 4 and 5, along with the symbols. For now, take notice of the matching shades of gray—dark, medium, and white. The intention is to use the darker levels to represent the bottom layers and the lighter layers to represent intermediate and top layers in each stack.

DESIGN CONSIDERATIONS AND CRITERIA

As with any project, content transport network design and construction is not successful without some amount of performance criteria. Content transport networks are different than ordinary voice and data networks. Ordinary voice and data networks don't typically do content transport well. On the other hand, design and build a network capable of transporting valuable program content, and voice and data can come along for the ride. Below is a list of considerations and criteria that can be used when preparing to undertake a content transport network project. The following list is not to be taken literally, nor is it exhaustive or all-inclusive:

- Access, switching, and transport elements
- Access and transport facilities can be terrestrial, satellite, or a combination
- Switch facilities will be time (TDM), cell (ATM), or packet (IP)
- Service availability is full-time 24/7 or shared
- Availability, reliability, robustness, grade, and quality of service
- Capital and operating cost
- Geographical or physical coverage includes local (LAN), metropolitan (MAN), regional, national, and global turf (WAN)
- LANs may have single or multiple segments covering a room, floor, building or group of buildings in a campus arrangement
- A MAN typically involves third party telco or ISP service and uses standard telephone facilities, such as E1/T1, E3/DS3
- WAN extends LAN and MAN to wider geographic areas not covered by local telephone companies and ISPs

Content transport networks can be built or bought, but practical realization is a combination of buying equipment and the rights to use facilities and services.

TABLE 3-1

Content	Transport technology or methodology					
	TDM	ATM CBR	ATM VBR	Frame Relay	IP	Ethernet
Voice (Real Time)	Yes	Yes	Maybe	No	No	No
Data	Yes	Yes	Yes	Yes	Yes	Yes
Video (Real Time)	Yes	Yes	Maybe	No	No	No
HTML Or Similar	Yes	Yes	Yes	Yes	Yes	Yes

Table 3-1 summarizes the types of network transport technology or methodology and it's usefulness for transporting various types of content.

Verbal descriptions of network elements—the terms used to deal with network elements—are a common source of confusion. Symbols used throughout the industry are equal sources of confusion. Who knows why it's that way; maybe this is nothing more than a set of behavior consistent with the many meanings and uses of the word *network*.

Adequately defining and describing a network is at best a challenge. Consider for example the famous, or infamous, cloud. The metaphor has more than a little in common with the white things in the sky. Clouds represent the weather. The weather is always changing, somewhat unpredictable, and never controllable. Networks are the same, except maybe a bit more controllable. In fact, the network operator or service provider should be in firm and complete control of any and all aspects of their network except things and effects they can't control such as the weather, errant backhoes, unpredictable motorists, and similar forces known to occasionally inflict themselves on power and telephone poles. In addition to the cloud, other symbols are used to denote different types of facilities.

Accordingly, a set of symbols should be adopted to define generic types of equipment and facilities to make block diagrams and topology maps. Table 3-2 shows some simple examples, most of which are used in this book.

TABLE 3-2

 The famous cloud that typically summarizes important details inside the network. Here, it will be filled with one of four shades of gray. The darkest layer corresponds to Layer 1. Layers 2 and 3 represent continuous and discontinuous bandwidth with Services Layer in white at the top.

 Layer 1 Point-to-point wired or wireless transmission facilities. Truly the "media layer" including electrical, radio or optical transmitters, receivers, repeaters, and add-drop multiplex. Arrows indicate full duplex path or route. Some call this level the "bit level" because it is in a form of continuous clocked bits

 Layer 1 cross-connect facility. Used to groom Layer 1 bandwidth into and out of Layer 2/3-switch and router equipment, and provision point-to-point private line facilities. Not to be confused with Layer 2 Circuit Switch, or other Layer 2 functions. In PDH networks, this layer is still considered the bit level, however framing makes it's way into the picture in the form of channelization built on individual 64Kbs timeslots.

 Layer 2 Circuit Switch represents switching required to support POTS and ISDN service. When other framing and protocols replace POTS or ISDN, a different type switch is required. For example, Ethernet or ATM.

 Layer 2 Cell Switch – structured, continuous bandwidth

 Layer 3 Cell Switch – unstructured, discontinuous bandwidth

 Layer 2/3 edge device; switch or router; structured, continuous bandwidth

 Layer 2/3 edge device; switch or router; unstructured, discontinuous bandwidth

 Layer 2/3 core device; switch or router; structured, continuous bandwidth

 Layer 2/3 core device; switch or router; unstructured, discontinuous bandwidth

Content transport network characteristics revolve around two basic types of content: live, real-time, or so-called steaming video; and non–real-time, such as realized when a file containing content is transferred across a network using *file transfer protocol* (FTP) techniques. Given these characteristics, then it seems logical to symbolize content transport with two types of network facilities and services.

Live, streaming content requires continuous, uninterrupted connections with an equal amount of bandwidth. That's the theory; however, in practice it's always prudent to leave just a tad of headroom. So how much is a tad? Practicality drives such in the form of how the service provider divides up the bandwidth and sells it. For example, a 10 Mbs ATM or IP network facility likely won't be precisely 10 Mbs. These animals usually break out in increments of octal numbers. So somewhere around 10 Mbs will be something like 10240000. If that is your choice of network transport channel, then the compression system output bitrate should be set at some number less than the channel rate. This parameter is also a victim of practical circumstances as well because these devices commonly have to deal with octal numbers. So a tad in practice happens to be the difference between the highest speed the encoder can be set at, and the channel rate. (See Appendix II for an example of calculating payloads and matching channel rates.)

Non–real-time content can be transported using continuous, uninterrupted connections, but it can also be carried on discontinuous bandwidth connections, usually at lower cost and improved utilization of the facilities. Be aware that realization of lower cost is dependent on obtaining use of facilities and services at unit prices based on time used and type of bandwidth occupied for each session or transmission just like the old fashioned long distance telephone call.

Standard network performance and characteristics must be understood before they can be applied to content transport networks. The next few paragraphs provide an introduction to time division multiplexing (TDM), ATM, and IP network technology.

TDM technology characteristics and performance are the standard cell and packet based network performance should be measured against. If a standard for TDM is required, then use wire, fiber, or another passive conductor of known performance. The characteristics of interest include available channel bandwidth, bit error rate, and jitter. However, in cell and packet networks, bit errors cause cell and packet loss or impairment, as can jitter.

ATM transport technology offers 5 classes of service. Constant bit rate (CBR), variable bit rate—real-time (VBR-rt), variable bitrate—non–real-time (VBR-nrt), unspecified bit rate, and available bit rate. While it may change in the future, ATM CBR is currently the only ATM class of service capable of transporting high-quality, high bit rate content in real time.

Packet-switched networks are inherently chaotic unless specifically configured to deal with continuous signal, or mixed-signal traffic and class-of-service. Packet networks are either Ethernet or IP. (Several packet or packet-like techniques exist; however, they only support content transport as a file transfer, not real time.)

In general there are two types of IP technology and methodology: Ethernet and Internet. The IEEE 802.1 standard defines Ethernet. Internet or more precisely, IP is defined in RFC 791. Ethernet transport of IP is defined in RFC894.

Ethernet architecture is built around shared media in the form of common set of cabling where the information is carried in packets, and the device such as a workstation or server listens or monitors the buss before attempting to establish a connection or session. The way the process works, end-to-end, has the sender and all the receivers constantly listening or monitoring the buss. A session is kicked off after a sender sends an initial transmission to all stations using a unique address. If the initial transmission has a valid destination address, that is an actual receiver connected to and listening to the buss, it responds with an acknowledgement. After the sender receives the acknowledgement, then and only then do the two computers establish a connection and carry on with the session using their unique address information.

IP networks, the Internet in particular, behave in similar fashion as Ethernet.

All these types of transport work well for moving files, including hypertext markup language—coded pages, fixed images, and other static objects. Uncongested networks may even support low volume continuous signals such as produced by voice or telephone service over IP, and even "work okay" with higher bandwidth continuous signals. Make no mistake about it though, unstructured networks cannot be relied on for transport of continuous signal, high bit rate, valuable content such as audio, video, closed captioning, control, or other signals associated with, or embedded in, program content.

Reliable, predictable, safe, and secure content transport requires network connections with sufficient bandwidth, grade, and quality of service (GOS, QOS). Even non–time-sensitive or non–real-time transport—so-called FTP—should be planned and implemented with care because of the size of the files and the time required to move them have significant economic implications.

Obtaining sufficient bandwidth, GOS, and QOS is a matter of specifying and configuring LAN, MAN, and WAN network resources.

Sufficient, continuous bandwidth means the network must exhibit bandwidth equal to or greater than the bandwidth of all traffic, not just program content if the network is required to accommodate email, web surfing, network management, and perhaps voice. Insufficient network bandwidth results in denial of service or, at best, delayed service. Program content payload bandwidth is roughly equivalent to the sum of compressed audio, video, and other signals multiplexed into a program stream or included in a file object stored on the system. When more than one real-time stream is present on the interface simultaneously, the aggregate of all the program streams cannot exceed the bandwidth available on the interface points of the sending and receiving systems and the network connecting the systems. In other words, the bandwidth of the sending and receiving systems must equal or, preferably, exceed the aggregate of all traffic.

GOS means the network connecting all workstations and servers must be available to all users within the design limits agreed to or promised to its users. For example, telephone network services use statistical probability based metrics to define and measure GOS level, inside and outside the network. A P.01 GOS means the network is designed and performs, or doesn't perform, within the limits of probability that the network will enable the user to complete the call in 99 of 100 attempts. This model can be applied to workstations, servers, and a LAN, MAN, WAN or combination of all and will perform satisfactorily 99 of 100 times when someone wants to transfer a file, or set up and use a connection to deliver streaming content originating on a server platform and terminating in one or more peer platforms at other locations or interfaces served by the network.

QOS means that the quality of the connection in terms of bandwidth, bit-error rate (BER), jitter, packet loss or any other parameter the payload may be sensitive to, is of sufficient level to support program content transport between and amongst the service points. The basic model for this category of network is classic TDM facilities found in ANSI/ITU standards-based networks. The acid test of performance is measurement and comparison to TDM private line facilities such as E1/T1, E3/DS3, and OC3/STM1. A good question of network equipment, facilities, and service suppliers is: Can you emulate T1, or DS3, etc.? The right answer is not "Yes." The right answer is, "You can expect jitter, packet loss and bit error rate performance of x, y, and z. This compares to TDM emulation performance of x, y, and z." Then you can decide if the differences fit into your required performance and compare one supplier to another.

Ingest, play out, and file transfer of program content as promised in many product and service descriptions require network connections with sufficient bandwidth, GOS, and QOS. Even non–time-sensitive or non–real-time transport—so-called FTP—should be planned and implemented with care because the size of the files and the time required to move them have significant economic implications.

Standard, so-called out-of-the-box or plug-and-play default LAN configuration included with recent generation Microsoft Operating systems (OS; W2000 Workstation & Server; WXP) enable non–real-time or FTP program content transport. Connect Ethernet to a network interface card (NIC) with access to the Internet, install the OS, run the Internet wizard, and voila! Instant success. No further fuss or effort and file transfer across the Internet from one host to another is possible.

These operating systems also permit configuration of an NIC to enable QOS as specified in IEEE 802.1p, a method whereby packets carrying continuous content can be marked and differentiated so LAN segments can isolate and protect the content from the effects of congestion and chaos mentioned above. Ethernet packets mapped to IP enable QOS marking to be passed to the IP network. If the network has differentiated services capability, real-time content transport across the network is possible. Older operating systems (NT 4; 95/98) do not include 802.1p/QOS capability.

Two types of connections are possible, and both may be required by the application. These include Unicast, or point-to-point, and multicast, or point-to-multipoint. These types of connections enable single or multiple deliveries of files or streams, sometimes referred to as *objects*.

The basic elements of a content transport network include customer premises equipment (CPE), access facilities at each location, and backbone transport in between. The equipment must be selected and configured to support the level and type of traffic. For example, if the traffic is program content only, that's one set of circumstances. If the network is to carry voice, data, and provide Internet access, that's another. If the network is to carry multiple types of traffic, the equipment and facilities will have to be structured to accommodate it. Figure 3-3 shows a general reference architecture capable of supporting voice, data, Internet access, and content transport.

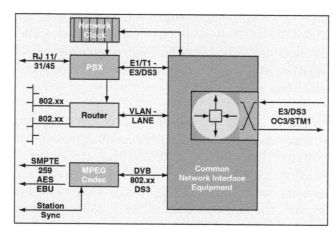

FIGURE 3-3   Premises Equipment Architecture

There are several characteristics of the architecture that should be pointed out and commented on. First, note the presence of a network clock reference and a separate station synchronizing reference. Neither has anything to do with the other and that's the point. The network clock reference is to make sure the network is stable and jitter-free because it must carry the embedded program clock reference along with the content. After all, if the network isn't capable of carrying the program clock reference to a satisfactory degree of accuracy, then the content will suffer impairment.

Although there appears to be a single-thread router and network interface, this is purely symbolic, and emblematic of the same level of redundancy as implied in the private branch exchange, LAN router, and Moving Picture Experts Group (MPEG) Codec. Resolving reliability, robustness and network performance concerns may require redundant equipment and facilities, with emphasis on content value and specific traffic levels. The terms and symbols are generic and intentionally chosen to cover several alternatives without stating them implicitly. For example, any new facility design should take a serious look at voice-over IP telephone service. New installations or even replacement/upgrade installations, may find economic advantage in fully integrated voice and data on LAN wiring. And although, not likely, it may be more appropriate to use ATM switching and transport for real-time program content than IP or TDM transport.

On the network side, there are similar issues and concerns; however, they must be addressed with carriers or service providers instead of manufacturers of equipment. As a design exercise, network access, transport, and switching should logically follow the food chain whereby the network facilities support movement of content within and between the creation, distribution, and delivery sections of the model. For example, moving raw, unedited content from a location to an editing facility, or moving finished program material from the post-production facility to a network operations center. And of course there's the end link, which requires the content to be moved from anywhere else to cable head end, DBS uplink, Internet access facility, or digital television transmitter input.

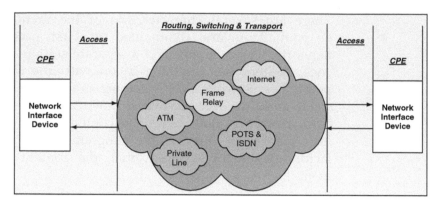

FIGURE 3-4   Reference Architecture

Figure 3-4 is a network topology diagram showing the details of how the basic elements fit into an overall architecture serving users located at separate sites, or operating centers.

All the various elements must be specified and priced out in detail. CPE is a capital investment. Access and backbone transport is an operating expense and can be provided by third parties, such as Internet service providers (ISPs), ILECs, CLECs, or inter-exchange carriers. Obviously, it is advantageous to deal with a single source for these services. Decisions on the end-to-end solution should only be made after following a due diligence process. Building the simplest of networks is not easy. Scaling start-up or small networks to larger networks becomes geometrically more complex. Churn and change after a network is built, debugged, and operational can be risky and should not be attempted without careful planning and deliberate, task oriented, sequential steps.

Similar to the end-to-end service model, the reference architecture simply lays out the functional components and shows how they relate. The NID or premises equipment interfaces and interoperates with the network to set up and tear down connections, monitor performance, and process alarms. The desired content transport network leverages one or more routing, switching and transport capabilities, depending on requirements and configuration of the access facilities. In situations where multiple types of traffic are converged onto a common access facility, the access facility must be channelized and mapped to the particular transport. For example, voice grade dialup or switched service would have channel capacity sufficient to accommodate peak voice demand on the public switched telephone network (PSTN) or integrated services digital (ISDN) network. However, if the design called for voice-over IP, some amount of bandwidth would be required to accommodate a similar level of voice traffic.

CPE is a router configured to connect to peer routers at the other locations. The router must be sized and have features selected to perform the functions required by the servers. These functions vary and depend on the encoded bit rate or payload of the content and level of traffic.

Another factor for careful consideration is distance between peer devices at other locations. If the distance is short, such as a nearby building, Ethernet could be an option. But outside adjacent buildings within a campus environment it's likely construction and capital cost will quickly add up to make third-party service providers with IP transport capability attractive.

Access facilities provide basic connectivity between the CPE and the MAN or WAN. Likely alternatives include E3/DS3 or OC3/STM1. Choosing an appropriately sized access facility is a matter of making a conservative estimate of initial traffic level, then monitoring the traffic and adjusting capacity to levels consistent with acceptable utilization and growth plans.

Backbone transport varies based on requirements and usually comes with significant and critical services attached. For example, the access facility is a dumb, point-to-point, TDM unchannelized facility. But routed networks include services such as routing and configuration protocols, IP address provision, service configuration and management that depend on processing functions resident in edge and core routers. It naturally follows that the owner of the core backbone equipment should provide these services.

LOCAL AREA OR LOCAL AREA NETWORK CONNECTION

LAN as used in this document means specific versions of IEEE 802.1, 802.3, and other standards applicable to network interfaces, referenced in a manufacturer's product documentation. Terms such as

10BT, 10/100BT, Fast Ethernet, Gigabit Ethernet, and 1000BT are informal references. Where doubt exists, it is recommended that designers rely on direct personal knowledge of formal IEEE or other applicable standards and seek clarification of applicability to a particular manufacturer's products from a representative knowledgeable about design details of the equipment.

Each standard Ethernet interface is assigned a unique 48-bit identification code by the original equipment manufacturer pursuant to a registry agent defined in IEEE standards. This code is referred to as a *MAC address* (meaning Media Access Control). The MAC address is inserted into Ethernet packets and enables devices connected to the LAN to establish sessions and route packets around or outside the LAN. When packets are routed outside the LAN, it's likely the MAC address is translated to an IP address, which becomes the basis for IP network transport. MAC and IP addresses are the functional equivalent of telephone numbers.

Ethernet interfaces enable user access, and connectivity between system components and peer platforms. The minimum required configuration is a simple Ethernet LAN segment. In more complex arrangements, the LAN segment provides access to MAN and WAN network facilities and services. If the network facilities are designed and implemented properly, use of any application on the system by any user with network access and user privileges is possible.

Most operating systems including LAN interface drivers permit manual or automatic, so-called plug-and-play configuration of the network interface. Microsoft Windows 2000 Workstation and Server and Windows XP Professional workstation include real-time communications client software. If the NIC is capable of being configured for QOS, then end-to-end real-time content transport is possible if it is configured. This is the first step in gaining end-to-end real-time content transport across the network between server platforms.

METROPOLITAN AREA NETWORK

Depending on initial size and growth plans, a local ISP may not be the best choice. For example if the requirement is for T1 access to the

Internet, then a local ISP may be appropriate. However, if the requirement goes beyond approximately 4 to 6 T1s, then it will make economic sense to use DS3. At this level, getting direct access to a Tier 1 backbone service provider starts to become attractive, economically and operationally. A Tier 1 backbone service provider is the highest and largest in terms of Internet hierarchy and coverage. Examples include AT&T, MCI, and Sprint in the United States. Similar networks exist in other countries as well. The point is the local ISP arranges for access to these networks and then aggregates traffic and resells the service. Direct access to the Tier 1 network also improves end-to-end performance and avoids finger pointing.

Alternatively, the customer can purchase point-to-point private line facilities and connect the routers together in the desired configuration. These facilities may range all the way from single or multiple T1s to OC3/STM1 or greater. In some areas, dark fiber may be available and has been known to be very attractive and cost effective. The down side, and many times the show-stopper, is network management, which is expensive in terms of capital and operating cost.

WIDE AREA NETWORK

WAN architecture is almost a duplicate of MAN architecture. The network equipment must connect to a local access facility, which connects CPE edge routers to network backbone routers. In general, common sense consideration based on physical geography can easily determine direction for a preliminary study. If for example, the customer sites are national in scope, then competing national, Tier 1 service providers works. On the other hand, if the sites are all local or state-based, local service providers—ISP, CLEC, or ILEC—can compete. Once a direction is set and a short list of suppliers is developed, final selection criteria will gravitate around capabilities, connectivity, and operational reputation.

CONTENT TRANSPORT NETWORK ELEMENTS

The approach to defining a content transport network is based on one of three conditions. Either the network exists and is not defined to suit the owner, or the network doesn't exist and the objective is to design, install, and operate one from scratch. The third situation is somewhere in between and the network will be replaced, downsized or upgraded to include more functions, capabilities and/or service sites. Ultimately, the content transport network contains approximately three or four basic elements: equipment used to make up the interface between the broadcast facility and the network, network access and transport facilities, and if the network is highly integrated with the broadcast operation, that is it becomes an operational function to change the configuration of the network, network management becomes the fourth element.

4

NETWORK TECHNOLOGY AND METHODOLOGY

Evolution or progression of any given technology is very often a product of extending currently understood methods to new and more effective capabilities rather than break-through inventions. Once technology is available, methodology evolves and progresses as well. It's quite safe (mundane too) but logical to conclude that methodology doesn't improve until technology is available for mankind to figure out how to improve the way it's used. Sometimes technology can't be well understood until after it can be used or experienced.

Simply cupping the hands together around the mouth and shouting results in directing sound through the air in a more focused way, increasing the distance of travel. Alexander Bell connected a concoction of carbon in series with another and a battery and wire to extend the distance sound could be carried. This is another example of how technology and methods continually evolve from the imaginations of some to meet the needs and wants of many. Modern communications networks—those dependent on electrical current and voltage behavior—have evolved over longer than 100 years. Networks are made up of discrete elements. Each of them has one or more technologies embedded in them. There's an economic food chain that comes into play as "gee whiz" technology matures and becomes commercially viable hardware and software products. Networks wouldn't be networks without these products. The products would not exist without these technologies. Today's networks are made up of many technologies embedded in tens of thousands of products, perhaps hundreds of thousands of products, if software is included.

This chapter continues the practical spirit of the book by dealing with technology in terms of how it's used. The time-honored phrase "practice makes perfect" could not be more useful, beneficial, or efficient when attempting to determine if technology is real or imagined, or if it's software or vaporware.

HISTORICAL BACKGROUND SUMMARY

Between around 1960 and 1980, the public switched telephone network underwent rapid and dramatic change from developments in solid-state digital technology. Initially, the diode and transistor were single function devices, but it didn't take long for them to be packaged into containers and branded *integrated circuits*. Computers—large, slow data processing machines and systems—were not immune to the same technological turmoil. Consequently, computers and their terminals migrated across the scientific landscape into office territory. Connections between the computers changed significantly as Teletype machine controllers turned into timeshare terminals. Someone figured out a way to convert the digital signal between the timeshare terminal and the computer from digital to analog, and reverse the process at the other end; devices made with modulator and demodulator techniques extended acronym territory with the term *MODEM*. All of a sudden the analog telephone network could connect timeshare terminals and computers as well as the Teletype network could. Originally, telephone networks were analog. Modems allowed telephone networks to be used to support computer communications.

As this initial impact from transistors and integrated circuit electronics enabled faster and faster computers, it had a similar effect on network technology. Bell Labs started working on digital transmission technology in the 1960s. The objective was to double voice channel capacity of a single trunk line from 12 simultaneous conversations to 24. This technology had tremendous value in large cities where the potential return was superior compared to digging up the street and burying more conduit.

Throughout the 1970s and 1980s, the long distance switching and transmission network underwent a conversion from analog to digital. Mini-computers replaced many mainframes; mainframes became faster and computer traffic grew. Data communications became full-time jobs for communications-savvy engineers and technicians.

A significant computer standard, developed in the late 1970s, remains in wide use today. The open systems interconnect (OSI) stack[1] defines a hardware section beneath a software section in a total of seven layers, bottom to top. The OSI stack makes a good framework for communications networks, including the Internet. Figure 4-1 shows the two-section, seven-layer stack with a brief explanation about what it represents and how it is applied. (see the section on "Layering as Used in Computer and Communication Networks," Chapter 5.)

Later in this and other chapters you'll find references to the OSI stack.

When the OSI stack was introduced, computers were just beginning to change from stand-alone islands into distributed processing

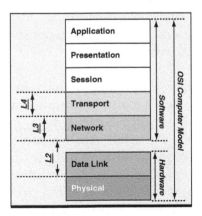

FIGURE 4-1   OSI Stack: Hardware and Software Sections

systems connected by data networks. The basic idea behind the stack concept is that each layer interfaces and interacts or communicates with the one immediately above and below, except, of course, the bottom and top layers for obvious reasons. If each layer successfully accomplishes its functions, then the system it's applied to should operate top to bottom. Attempts to map Internet and Telecom functions or processes to the stack are made from time to time, but in isolated ways such as a reference to *layer 2 switching*, or *layer 3 routing*, or even *layer 2/3 switching or routing*. These references seem to be more of a way to characterize a particular switching or routing function in terms of the OSI stack, rather than applying the OSI stack to communications networks in general. Furthermore, it would seem to be useful in analyzing and structuring or designing networks capable of carrying disparate, converged traffic types on a common access or transport facility.

BASIC NETWORK ELEMENTS AND FUNCTIONS

Communication network architecture (yes, including the Internet) includes six critical functional elements, or capabilities: clocking, multiplexing, routing, signaling, switching, and transmission.

Clocks control basic timing in digital networks. Digital networks simply wouldn't work without accurate, consistent, long-term, stable clocking and timing mechanisms. The basic clocking scheme used to maintain timing and synchronization in networks is not much different than it was when first conceived in the 1950s, except it's significantly more accurate and much less expensive, especially at the higher levels of accuracy and precision.

Multiplexing enables two or more signals to share time and/or bandwidth of a common facility. Multiplexing gains greater use of a limited resource. Multiplexing was a key characteristic of early analog telephone systems. Analog multiplexing shares frequency spectrum instead of time. Multiplexing can be active or passive. Active multiplexing involves electronic circuitry, while passive multiplexing, sometimes referred to as combining and filtering, requires no power supply, and attenuates the signals being combined.

Demultiplexing simply reverses the multiplex process. The multiplexing techniques used in classical T-carrier networks are active at the bit level. Timing differences between signals generated by disparate clocks running within frequency tolerance specification limits, along with a variation in propagation delay of the transmission path require the use of bit stuffing techniques to avoid clock slips and errors in transmission.

Routing in its broadest context applies to multiple ways to get from here to there, or connect point A to point B. A *router* or *routing switcher* in a broadcast facility is a drastically different beast than a router that can pass Internet packets from one port to another. Routing telephone calls and configuring private line connections play an important part in the global communications network today and are likely to remain so well into the future.

Signaling is the mechanism whereby customers, subscribers, and users (through equipment) communicate with the network to setup and tear down a connection, or configure it for initial use, or reconfigure it for different use (i.e., change the default service configuration). Signaling is also a process whereby network elements communicate with each other in response to commands from users for service, or the owner for changes in configuration or service capability. Successful signaling depends on a logical addressing or numbering scheme whereby all the elements in and outside the network carry a unique identification label.

Switching has been around since someone had a hunch that telephone service could take a cue from the railroads and get more use from fewer telephone lines by installing a switching point somewhere in the service area. From automatic switch-over when a transmission backbone segment fails, to provisioning private lease lines, to telephone service, data communications, audio and video conferencing, content creation, distribution, and delivery, modern communications networks simply wouldn't do what they do so well without it. Switching concepts include circuit switching, cell switching, and packet switching.

Transmission is the act of propagating energy or moving information from point A to point B. In the context of communications networking, the term includes sending and receiving. If the heart of the network is the clocking system, transmission is analogous to the arteries and capillaries carrying oxygen from the lungs to the brain and other important organs. Modern communications network transmission seems to have started when someone figured out that a direct current voltage applied to one end of a pair of wires could be detected at the other end as long as the conductive characteristics of the path are intact. Without the underlying transmission facilities, today's IP would be of no more value than Samuel B. Morse's telegraph code without a baseband electrical signal transmission facility. Successful transmission requires a viable medium. Electrical transmission works well on copper wire. Radio transmission moves easily through free space, where electrical current doesn't travel well. Light waves move through transparent glass, but opaque objects block them.

CLOCKING, TIMING, AND SYNCHRONIZATION

Regardless of the type of transmission media (wire, radio, or fiber), bits are sent and bits are received. Every receive port in the network has to deal with an incoming serial bit stream that includes clocking and payload bits. Clocking and data, also called *payload*, have their timing and phasing relationships established at the point of creation. Along the way, the serial bit stream may be multiplexed with additional serial bit streams, cross-connected to a different carrier, or switched at the circuit, cell, or packet layer. Yes, this is layer 1 and 2 of the OSI stack. Separating clock information from payload, or mucking around with the time relationship between signal transitions amounts to errors. If, for whatever reason, a network element loses its synchronization reference and wanders outside holding limits, anything and everything using it as a synchronization reference is out of time with the larger network, converting valuable data to invaluable trash.

Clocking and data recovery are a critical function. Considering the clocking concept from the perspective of a receive port on a network element or the receive end of a transmission path gives the ability to look backward toward the source and forward toward other network elements and facilities dependent on the clock for proper operation and delivery of the associated payload.

If there is a single point in the entire end-to-end, top-to-bottom process that is the most critical in moving digitized information through a network, it has to be at each and every receive point in the network. Receiving the bitstream intact and then extracting clocking information must be done before the payload can be extracted. Even if the receiver is clocked externally, clock and data signals in the incoming stream must be separated and defined. Payload framing depends on timing relationships, and guess what, timing depends on clocking. Serial data streams just aren't real if they don't have a mechanism delineating framed data and clock signals. Understanding clock and data recovery as a stand-alone function is one of the keys to understanding digital communications networks. Figure 4-2 is a block diagram of a typical clock and data recovery function found in almost every network element.

The functions in Figure 4-2 include receiving light or radio waves from the media, detecting and converting it to electrical voltage transitions. This requires a lens or optical interface in the case of light waves. The equivalent in radio waves would be an antenna. The purpose of either is to detect and focus the received energy into a signal containing the time variations between and during presence and absence of signal. The signal is then passed to a detector, which converts the signal into a series of transitions representing the serial bitstream originally sent by the transmitter.

The raw bitstream is then fed to a clock extraction circuit and a buffer. The clock is extracted and used for two purposes: one is a clock signal representing the original clock used when the data was created or multiplexed, and the other drives a data recovery circuit. The data recovery circuit extracts and re-clocks the data to restore as precisely as possible the transitions between 1s and 0s making up the data. At

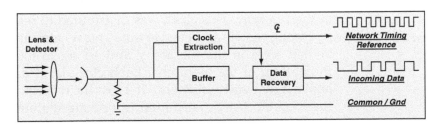

FIGURE 4-2 Clock and Data Recovery Functional Block Diagram

this point, the data is simply just that. We don't know anything about framing or channelization and nothing about protocols, byte size, packing, or error correction. These are all to be determined with additional processing, and take place through the upper layers of the ISO stack.

The clocking signal is representative of the original clock sent by the transmitter. It also has a lot of unknowns, but the important point to recognize is that this signal represents timing from another node or remote site outside the physical boundaries of the receiving equipment site. This signal could be important in the overall scheme of things or it could be irrelevant. For example, it could be used as a timing reference for the entire site. If that were the case, then many other considerations become important. If it is used as a timing reference for the entire site, there is no question that it will continue being proliferated to other sites because it becomes the clock that turns up as the receive clock after clock and data recovery at other sites receive signals from the instant site.

Naturally, this begs a few questions: "Where did that clocking signal originate?" "What is its level of accuracy?" "Is it a network clock?" Clocking in communications networks is as critical and necessary as synchronizing and time code in television audio and video. Timing in communications networks can be as complicated or as simple as timing in digital audio and video systems. Clock accuracy in communications networks is built around a four-level hierarchy with the most accurate clock at the first level and least accurate at the fourth level. Originally developed by Bell Labs for AT&T's digital network in the 1960s, adopted and standardized by ANSI, ITU, and other standards-making bodies, the system is referred to as *stratum 1* through *stratum 4*. In the beginning, there was only one stratum 1 clock in operation for the entire network at any given time. This clock was distributed to regional switching and operating centers and used to time and synchronize network elements at that level in the hierarchy and so on to local central offices at the bottom of the hierarchy. Each clock level depended on the one above it for synchronization. If it lost the reference signal from above, the local clock was permitted to operate within a range wider than the higher reference until the higher-level reference was once again available.

As you can imagine, there's not much of a disturbance when the reference clock disappears. But what about when the reference clock returns or if it is intermittent? Dealing with these situations led to the development of a *hold-over* specification, which basically requires the clock to remain within, or hold its frequency of operation for a given time and, in the interest of overall stability, not switch back to the higher hierarchy reference clock until after some period of stable operation, and if possible, switch gracefully, sometimes called *hitless switching*.

Primary reference clock (PRC) products compliant with ANSI and ITU standards are available from many sources. These clock sources can run independently, or they can be referenced and locked to other sources traceable to the World Timing Standard for time-of-day, called Coordinated Universal Time (UTC).[2] UTC is the result of combining TAI (International Atomic Time) and Universal Time 1 (UT1). TAI is a timing reference derived by averaging outputs from the clocks of approximately 100 countries. These clocks keep the timing relationship to each other within 2 to 3 millionths of a second over a year. UT1 provides a correction to compensate for the difference in solar time and TAI caused by slightly elliptical orbit and polar inclination, both of which affect solar time.

Distributing the PRC to all the elements in early digital network infrastructure was expensive, cumbersome, and complex. However, advances in lowering the cost of clock reference products and the availability of the global positioning system (GPS) and Internet-based references running under network time protocol (NTP) have dramatically reduced cost on all fronts and improved reliability and accuracy of the sources and their references.

Why all the fuss and bother? Essentially, signals derived from multiplexing low-speed digital bit streams into higher-order aggregate bit streams, some of which provide private line, and others that provide circuit-, cell-, and packet-switched services must be synchronized and timed exactly the way digital audio and video signals are timed and synchronized, including SMPTE time code, to bit level accuracy within a frame. The only differences are that Telecom digital signals aren't subject to switching transitions such as a split screen or cross-fade, and telecom and digital program content run on entirely different time bases. One other point is that this timing accuracy has absolutely nothing to do with the transmission media—it makes no difference if the transmission is satellite or terrestrial radio, optical transmission, or baseband electrical signal transport.

One thing to pay attention to in systems where the bit stream is modulated onto a carrier is the capability of a particular modulator to lock the carrier frequency generator to an external source such as the incoming digital signal or PRC. On the receive end, it may be appropriate to lock the local receiver beat frequency oscillator in the receiver to either a PRC or the incoming digital signal clock. Including these capabilities in a piece of equipment is not without cost; however, those that do provide that capability will be capable of better bit error performance across the transmission facility because of the absence of asynchronous cross talk causing clock slips.

Lastly, be careful not to confuse network primary reference clock with program clock reference in Moving Picture Experts Group (MPEG). The two are completely different animals and have nothing to do with each other. When MPEG2 program streams are transported through networks, the PCR is just another set of bits in the payload. Clock distribution and network synchronization is covered in detail in Section 5 of Chapter 5, entitled "Clocking Considerations" as part of overall network architecture.

MULTIPLEXING AND FRAMING

Another key to understanding digital communications networks is multiplexing and framing. If you understand layer 1 clocking and data recovery, then clocking and data recovery in layer 2 will come naturally. Many things happen at layer 2, but the overall function is to keep the bits organized—framed, timed, and synchronized. Lacking that, the higher layer functions don't do very well because they assume these functions to be intact and functioning at the lower layer. Put another way, clocking and synchronization are only implied at layer 3 and above. Two major multiplexing and framing approaches in current use include T-carrier and synchronous optical network/synchronous digital hierarchy (SONET/SDH).

Communications network-multiplexing techniques grew out of the T-carrier concept created as a family of trunking techniques. Dubbed *pair gain*, the goal was to double trunk capacity from 12 to 24 channels or talk paths on a single trunk facility. In a few short years, the technology morphed its way into the network architecture. Its key parameters include 8-kHz sampling and 8-bit quantization. This results in a single voice channel rate of 64 Kbs. Sampling 24 voice signals simultaneously, or within the time frame of one sample, results in a framing time of 125 microseconds. When these 24 channels, or DS0s, are bundled into a serial stream with a single framing bit, it aggregates to a T-carrier at 1.544 Mbs. Figure 4-3 shows the duration and timing parameters of a channelized T1 or DS1 signal.

Twenty-four clock samples of 8 bits each result in 192 bits of payload. After each 192-bit sample is completed, it is serialized and placed immediately following a framing bit making a 193-bit frame. The resulting frame rate is maintained at the same 8-kHz clock rate, resulting in a 1,544,000 bps stream. The 1,544,000 stream transports 1,536,000 payload bits, or 192,000 payload bytes and 8000 framing bits, or 1000 bytes. Each individual 64-Kbs channel is referred to as DS0, meaning digital signal level 0. The composite serial bitstream is referred to as digital signal 1 or DS1. DS1 is considered a multiplex level and while the bitrate and framing is the same, DS1 is often confused with the T1 carrier or channel and its interface. See Table 4-1 for examples of the different multiplexing level and transmission channel interface rates.

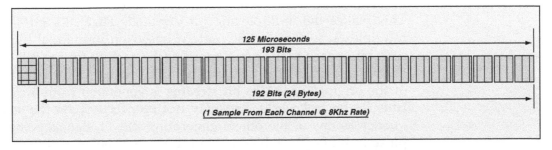

125 Microseconds
193 Bits

192 Bits (24 Bytes)

(1 Sample From Each Channel @ 8Khz Rate)

FIGURE 4-3 T-Carrier Digital Channel Timing and Framing

TABLE 4-1

Channel interface	Signal level	Multiplex input	Payload Channel count	Bitrate	Clear channel rate*	Aggregate multiplex rate	Region of use
T1/J1	DS1	24×DS0	24	1.536	1.536	1.544	NA & Japan
	DS1C	2×DS1	48	3.072		3.152	E Countries
	DS2	4XDS1	96	6.144		6.312	E Countries
E1	E1	30×DS0	30	1.920	1.920	2.048	E Countries
	E2	2XE1	120	7.680		8.448	E Countries
E3	E3	16XE1	480	30.720	30.720	34.368	E Countries
T3/J3	DS3	28XDS1	672	41.216	43.008	44.736	NA & Japan
	E4	64XE1	1920	115.200		139.264	E Countries
	DS4	6×DS3	4032	258.048		274.176	NA & Japan

*The minimum amount of bandwidth available. It is also precisely equal to the number of 64 Kbs DS0 channels removed from the next higher order channel. Actual payload in practice is dependent on design requirements. For example, a DS3 requires a minimum of 526,306 bits for framing, leaving 44,209,694 out of 44,736,000 bits available for payload in unchannelized service.

BASIC FRAMING

Framing bits are also applied when a basic 1.544 Mbs stream is multiplexed with another stream or additional streams into higher order aggregate signal. Additional bits are added in specific timeslots and designated as framing bits to enable the receiving equipment to recover the original clock and separate the payload, first into the next lower order bitstream and then ultimately down through the multiplex hierarchy to the original 1.544 Mbs payload and 24 individual 64 Kbs voice channels or DS0 signals.

This makes the multiplexing bit oriented. That is, each stream is multiplexed into a specific pattern based on individual bits where each bit in each frame has a specific (theoretical) timing relationship to the same timeslot in peer bit streams. Because of the fact that each of the original 1.544 Mbs bit streams is generated from a clock that runs in the real world, and may not be precisely on the same frequency as any of the others generating the T1 signals being multiplexed, and because the timing of the signals being multiplexed may change due to propagation delay variation in the transmission media, the resulting aggregate signal multiplexing is said to be *plesiochro-nous*, meaning almost or nearly synchronous, but not asynchronous or non-synchronous. Multiplexing of signals from disparate clocks that are almost or nearly synchronous requires another technique called bit stuffing.

Bit stuffing[3] is exactly what it's name implies, adding or "stuffing" bits into a multiplexed stream to raise the speed, or number of bits per unit of time, so there are enough bits to fill the timeslots in the higher order channel. For example, when 4 DSI signals are multi-plexed to make up a DS2 signal, one of the signals is sent at exactly, 1,544,000Bps, one has 364Bps, the third gets 314Bps, while the fourth gets 414Bps, making an aggregate for the DS2 of 6,306,272Bps.

When 7 DS2 signals are used to build up a DS3, still more bits are added to each DS2 to enable the network to accommodate the disparate nature of the various DSI clocks and multiplexing oper-ation.

The T-carrier concept originated in the United States, but was followed in due course in other countries. Initially designed for four-wire media, it found its way to coaxial cable, wireless, and optical fiber media. An international version of the DS1 is called an E1. It uses the same 8 Kbs sampling and 64 Kbs DS-0 channel rates, and 125-microsecond framing. However, 30 timeslots are placed in the 125-microsecond frame, resulting in a payload of 240 bits per frame, a payload rate of 1.920 Mbs, and a total channel rate of 2.048 Mbs.

Another major difference and significant improvement of E1 over T1 structure is increased overhead. From the start, this was a troubling characteristic of T1, not because it was too much, but because it wasn't enough. There was never a standard method, nor enough bits to deal with the many overhead requirements for voice service. Besides, when data transport came on the scene, US DS0 channels could reliably deal with only 56 Kbs instead of the entire 64 Kbs bandwidth. So the designers of E1 digital facilities added 2 to 64 Kbs channels providing 128 Kbs. Adding these two timeslots in the 125-microsecond frame resulted in 256 bits in each frame, 240 for payload and 16 for overhead.

CHANNEL RATE VS. MULTIPLEX RATE

One of the major points of confusion, both inside and outside the Telecom industry is the misunderstanding and misuse of channel interface and multiplex rate terms. In publication after publication, the PDH hierarchy is explained by simply calling out the various multiplex levels or signal rates without delineating the difference between the rates that specify channel rates through the network, and the rates resulting from multiplexing.

Channel rates apply to channel interface, which is the point of network access and egress. Table 4-1 shows most of the PDH multiplex aggregate rates and channel rates covered by global and national standards. Basically, the standards are structured toward use in three regions: Europe and the rest of world (E), except North America (T) and Japan (J). Multiplex levels generally are designated with a DS, meaning digital signal (level) and channel interface and bandwidth with a T, E, or J.

Another factor that is often completely missed is the fact that telecom networks are dominated by voice traffic. The nature of voice traffic caused the network to evolve into circuit switched 64 Kbs channels. These channels are multiplexed into higher order bit streams. An OC192 transmission facility configured to handle only voice traffic contains 192-DS3 equivalents, each capable of 672 64 Kbs channels for a total of 129,024 unique, independent segments of bandwidth. An STM64 configured to handle only voice traffic contains 122,880 bandwidth segments. What about using some of this bandwidth to carry something of a different nature than a 64 Kbs voice? It is entirely possible, and that's where network channel interface comes into play. From a network transmission facility perspective, a bit is a bit. The transmission network doesn't know or care if the bit carries voice, data, or anything else, including errors, as long as it meets network clocking and time duration parameters. What this means is that the transmission network will accept access and egress or drop and insert segments of bandwidth that just happen to be sized and structured to fit in place of 24 (T1), 32 (E1), 480 (E3), 672 (T3) voice grade equivalent, or 64 Kbs channels.

Another point of confusion is unchannelized or clear channel rates. This type facility is used for non-voice channel applications such as

Internet access, wholesale encrypted traffic (encryption applied to a T1 or group of individual channels rather than individually encrypting each channel), asynchronous transfer mode (ATM) network, physical convergence layer protocol (PLCP) access, compressed program content over E3 or DS3, and other applications that require contiguous bandwidth that cannot operate over 64 Kbs channelized facilities. IP routers are likely to have a PDH wide access network interface at T1/E1/J1 or E3/J3/DS3.

SYNCHRONOUS OPTICAL NETWORK/SYNCHRONOUS DIGITAL HIERARCHY MULTIPLEXING

SONET/SDH multiplexing techniques are built around a fixed size row and column structure that contains overhead (OH) and a synchronous payload envelope (SPE). The SPE is designed to carry synchronous transport streams (STS). The dimensions of the SONET/SDH building block are shown in Table 4-2.

SONET/SDH multiplexing frame structure is shown in Figure 4-4. Notice the structure is built around the same 125-microsecond, 8-kHz timing and synchronization structure as PDH multiplexing and framing. The difference however, is that STS really is synchronous—not asynchronous, plesiochronous, or non-synchronous—because unlike the others, SONET/SDH clocking is orders of magnitude

TABLE 4-2

Frame Duration:	125 Microseconds	
Frame Repetition Rate:	8,000 Frames/Second	
Synchronous Transport Stream (STS−1) Bit Rate:	51.84Mbs (6.48MB/Octets/second)	

Overhead	Columns	Rows
Section	3	3
Line	3	6
Path	1	9
Sub-Total	4	9
Payload:	86	9
Total STS-1:	90	9

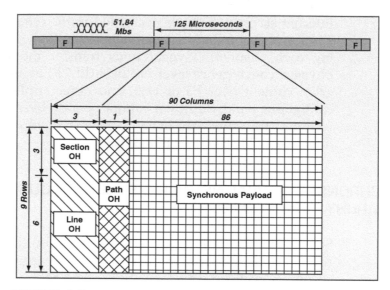

FIGURE 4-4   SONET/SDH STS-1 Multiplex Frame Structure

more accurate than PDH clocking, and multiplexing is byte-interleaved instead of bit-interleaved.

Think of the intersection of each row and column as defining a single byte. Given 8 kHz and 125-microsecond framing, the basic STS bit rate calculated as follows:

Overhead and the associated envelope and transport stream are fixed in size and duration, resulting in a fixed bitrate. Overhead is further subdivided into channels and functions used to configure, manage, monitor, and control the line and section elements of the link. Another overhead is path oriented and provides end-to-end management and monitoring capability across tandem connections passing through terminals and extending to customer premises when an STS facility is built from multiple links.

The SONET/SDH structure is extendable from one stream to 192 streams. Each STS consists of 86 columns by nine rows or bytes of payload and one column by nine rows of bytes for path overhead. Section and line overhead are unique to each link. Path overhead extends across all links in a path made up of multiple STS passed through a terminal intact.

An OC3/STM1 (155 Mbs) SPE includes three STS1 transport streams. Each STS1 stream can accommodate one or more virtual tributaries (VT). A VT is used to transport a single or as many as seven PDH signals such as DS1, E1, or DS2. An asynchronous DS3 is transported on a single STS1.

In SDH, the equivalent to the VT is a tributary unit (TU) at the lower bitrates and virtual container (VC) at the higher bit rates. For example, an E3 or E4 would be mapped to a VC. Figure 4-2 shows a partial list of channel interfaces and multiplex rates. Like PDH multiplexing, it was crafted to serve fixed bitrate applications requiring 64 Kbs increments. However, it is possible to map other synchronous signals such as ATM, HDLC (high level data link control), SDLC (synchronous data link control), and PPP (point-to-point protocol) to SONET/SDH without going through the PDH VT/TU/VC layer. Table 4-3 shows the more common PDH, SONET/SDH interface, and channel rates as well as their relationship to each other.

TABLE 4-3

PDH			ANSI-SONET			ITU-SDH		
Channel interface	DSO count	Mux rate	Mux layer	Line rate	Line layer	Mux layer	Line rate	Line layer
T1	24	1.544	VT-1.5			TU-11	1.544	
E1	30	2.048	VT-2			TU-12	2.048	
	48	3.152	VT-3					
	96	6.312	VT-6			TU-2	6.312	
	120	8.448				TU-2	8.448	
E3	480	34.368				VC-3	34.368	
T3	672	44.736	STS-1			VC-3		
			STS-1	51.840	OC-1		51.840	STM-0
E4	1920	139.264	STS-3	155.520	OC-3	VC-4	155.520	
			STS-3	155.520	OC-3		155.520	STM-1
			STS-12	622.08	OC-12		622.08	STM-4
			STS-48	2488.320	OC-48		2488.320	STM-16
			STS-192	9953.280	OC-192		9953.280	STM-64

MOVING PICTURE EXPERTS GROUP TRANSPORT MULTIPLEXING

Another name for the multiplexed aggregate bit stream is *transport stream* passed to a standard network interface, transported to a distant location, and then demultiplexed into the original streams. Figure 4-5 shows how audio, video, and data (elementary) streams are aggregated into a transport stream in MPEG.

In MPEG multiplexing, elementary streams are multiplexed into a higher order bit stream similar to that done in communications networks. Multiple program transport streams can in turn be multiplexed into higher order streams. In some cases statistical multiplexing techniques may be used to allocate unused aggregate bits to each of the program sources, depending on activity.

The main point to make here is that the connection between the multiplex and demultiplex can be any standard communications channel capable of interfacing the end units and providing the band width, class, and quality of service required by the television signals being transported. Note that in this depiction, audio, video, and data are arbitrarily shown with one-way paths. These could easily be two-way, requiring full-duplex network facilities. Note also the 10/100BT access requires full-duplex layer 2 as well.

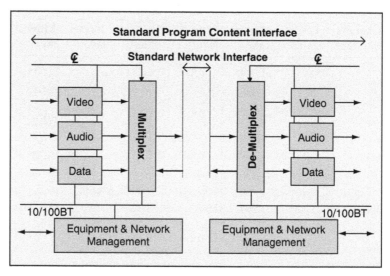

FIGURE 4-5   Program Content Multiplex and Demultiplex Concept

ROUTING

Routing and switching mean different things to different people. For example, both terms apply to switching and transmission facilities. Routing in the circuit switched or voice services world means that a call is routed according to service configuration parameters in a PBX or end office switch. For example, least-cost routing is established when the originating switch is programmed to use the least expensive route between the origination point and termination point for a telephone call. Alternate routes may be a point-to-point private line, a virtual private network (VPN), or the public network, where the private line is the lowest and fixed cost, the VPN is the next least expensive, and the public network is the most expensive.

Wholesale routing of traffic occurs when the traffic is moved from one transmission facility to another. For example, traffic normally routed from New York to Atlanta may go direct, but an alternate route may pass through Cincinnati. A national fiber ring might have a southern route and a northern route, as a regional ring might have an eastern route and a western route. Traffic normally on an Intelsat transponder facility might be moved to another route using a transponder on a PanAmSat satellite.

In the data world, routing becomes more of a technical issue because of the underlying network technology. From a classical perspective, data networks were built using leased or private line facilities provided by carriers on a 24/7 basis at fixed prices. For an enterprise with a headquarters located data center and several field operations, private or leased lines are used to connect computer terminals to the data center. Depending on the number and geographic locations, all the field offices may be connected directly to the data center in a home run arrangement. However, if two or three of the remote operations were physically close to one another, it may make sense to hub them into one common location, aggregate traffic, and connect to the data center over a common facility. In this case the hub becomes a routing point. All traffic from the other nearby locations is routed through the hub to the data center.

Data communications routing can become very complex and confusing because of the proliferation of various flavors of ATM, Ethernet, frame relay, and IP techniques, to say nothing of classic SNA, X.25, HDLC/SDLC, BISYNC, and others.

In circuit switching, routing intelligence is in the user's head and the network, and is used to tell the network how to route a call or make a connection. The same basic principle applies to ATM, frame relay, and IP networks as well. In the circuit switched network, the routing intelligence resides in the common channel interoffice signaling system and its configuration software. In ATM, the intelligence is included in each cell header and distributed across switching machine configurations. In Ethernet, frame, and IP networks, it's in the packet headers and distributed across configuration parameters in switching machines swooned over and lovingly called *routers*—except frame relay, which is typically a meshed, point-to-point arrangement and therefore has limited connection capability.

SIGNALING

There are two key parts to the signaling process: an identification scheme and a linking or connection process. Signaling can be initiated by the user or by the network. Signaling can also take place between two or more users using the network. Routing and switching depend on signaling within the framework of voice, data, and Internet.

All of the switching and transmission in the world wouldn't be worth much without signaling capability. Signaling has many forms and capabilities, but only one function, which is to enable the user and the network to communicate with each other. The user tells the network what to do. The network tells the user that it has done it or failed to do it. The simple model has been around since dial telephones replaced human operators. Successful signaling and network operation requires a unique terminal or telephone numbering and identification plan whereby each unit or telephone is the only unit on the network with that number (Table 4-4). The equivalent in a local area network (LAN) is the media access control (MAC) address. In the Internet, it's the IP address.

In North America, the telephone numbering scheme relies on a fixed, 10-digit numbering plan. Outside North America, the number of digits in the plan will vary from country to country or administration. Each administration or country relies on a country code. These numbering plans were conceived and are maintained according to ITU recommendations. So the telephone number plan, or identification scheme, is service provider-based.

In the world of the LAN, however, the identification of each computer or host is manufacturer-based. Each network interface card (NIC) is given a unique 48-byte identification number called a physical address because it is put there by the manufacturer as part of the hardware manufacturing process and remains fixed in the card for life. The IEEE administers the standard for creating and producing these numbers for each card. The process requires each manufacturer to register with the IEEE, whereupon they receive unique manufacturer identification, and then the manufacturer is responsible for creating unique numbers for each card built and shipped. To be just a bit more specific, a MAC address is assigned to each network port on any device. For example, many motherboards in computers have built in LAN ports. Other plug-in cards may be built to accommodate multiple LAN connections such as a virtual local area network (VLAN) cards in a router, or LAN emulation (LANE) cards in an ATM switch.

Internet identification is accomplished by assigning network-wide unique IP addresses to each device on the network. Successful connectivity between two or more hosts on an IP network involves a translation scheme where the unique IP address is subjected to an address resolution process involving the local network addressing scheme or protocol and the physical or MAC address. For example, the local network is likely to be some version of Ethernet.

TABLE 4-4

Network type	Application or use	Signaling technique
Data Networks	LAN, MAN, WAN Connectivity	Layer 2 IEEE 802.x Logical Link
Voice Networks	Local, Regional & Long Distance Calling	DTMF, CCS7/SS7
Internet	Network Connectivity	PPP, L2TP IGP, BGP

Signaling, even in general, is beyond the scope of this book. However, literature is widely available from many sources, the least expensive of which is your favorite Internet browser.

SWITCHING

Here we will consider several types of switching: circuit switching, cell switching, and packet switching. While there's less confusion surrounding switching than routing, there's still enough to elicit caution when discussing, reading, or writing about the subject. The confusion is not so much a matter of the function as it is the nature and character of what is actually being switched and the mechanics of the switching process taking place. Fundamentally, switching enables a full or part-time fixed bit rate transmission facility to be shared.

CROSS-CONNECT SWITCHING

Cross-connect switching is almost exclusively a telephone company term and function. It's nothing more than an electronic version of a manual patch panel. It is used to groom bandwidth and provide private or leased-line facilities. Its formal name is *digital access and cross connect switch* (it is also referred to as digital access and cross-connect system) and, as you might guess, it has been given acronyms: DCS and DACS.

There are several types of DCS. The earliest DCS simply enabled connection of a DS-0 segment of bandwidth or time slot from one DS-1 facility to another inside wire centers and central offices. As higher order multiplexing equipment became available, more DCS functionality appeared and enabled switching of not only DS-0 level circuits, but also whole DS-1 circuits across multiple DS-2 or DS-3 facilities.

SONET/SDH DCS has several capabilities, including grooming bandwidth at DS0 (64 Kbs) up through all the VT levels and the STS level (51 Mbs). For instance, it is possible to cross-connect a 64 Kbs segment from one VT into another VT on a different STS in the same or another

OC level. It is also possible to cross-connect an entire 51 Mbs STS in one lambda to a different lambda on another path or route.

CIRCUIT SWITCHING

Circuit switching and routing is the basis for all domestic and international telephone or voice grade, dial-up traffic. The circuit switching function is distributed between end-office switching systems and network switching systems. End office switching systems may be a private branch exchange (PBX) physically located on subscriber premises, or a partition in the telephone company's nearest office, commonly referred to as *centrex* service. Network switching systems include the local serving central office and any other systems facilitating a path for a telephone call. Nowadays, these systems range in size from a few thousand to hundreds of thousands of ports capable of handling millions of calls per hour.

Circuit switching in functional terms is nothing more than connection of one transmit-receive pair on one side of a switch to a transmit-receive pair on another port on the same path or route, or a different path or route, sometimes called the other side of the switch. Tandem switches are nothing more than transit points that link up network or inter-network transmission facilities. For example, each of the 200+ local access and transport areas (LATA) in the United States has a minimum of one tandem switch, which acts as the transit point between the access and transport networks used by long distance carriers to carry calls from one LATA to another.

Voice grade dial-up service is almost all digital in the United States. However, many analog switches remain in other parts of the world. Where digital switches provide the service, integrated services digital network (ISDN) services—really an access method, not a service—is available. In highly populated areas of many countries, digital subscriber line (DSL) access is available and growing.

Transmission bandwidth available in circuit switched facilities varies from below 64 Kbs (rarely more than 49 Kbs) to 1.536 Mbs. The limitation in analog service is a matter of the ability of a modem to talk to another modem over a local telephone loop. Of course, it doesn't much matter to voice grade service. After all most, if not all, telephone equipment is bandwidth limited to around 3.5 kHz, which fits easily into 8-kHz sampled PCM.

ISDN and DSL access provide higher capabilities though. ISDN Access is either 144 Kbs, called basic rate interface (BRI), or 1.544 Mbs, called primary rate interface (PRI). BRI is channelized into three channels, two bearer or B channels at 64 Kbs, and one delta or data or D channel (16 Kbs) used for signaling and control purposes. PRI access is facilitated with T1 transmission facilities and is channelized into 23 to 64 Kbs B channels and 1 to 64 Kbs D channel. It should be emphasized that the previous explanation is purely in terms of technical capability. Leveraging the bandwidth into variable amounts and getting charged for it on a case-by-case, service-by-service basis is an entirely different matter.

For example, ISDN-based Internet access never achieved large usage because the equipment used by ISPs and their users was limited to BRI rates—64 Kbs at best. And because the ISPs are not the telephone company and have no capability, such as a big digital circuit switch, and have no funds available to buy a big digital circuit switch and therefore no interest in competing with the telephone company, they only offer Internet access service. From the telephone company viewpoint, they simply are prevented from being in the data services—Internet access, or Internet service provider (ISP) business—by current FCC rules and legislation. The telephone company can only sell POTS, ISDN, or private line service. It cannot offer any type of switching other than these services. Some of the independent non–regional Bell operating telephone companies have purchased and operate ATM equipment, but basically they are quite limited in the service they can provide using these or other non-voice service, frame relay, and IP-based switching and routing systems.

DSL access varies according to several factors, the main one of which is the distance between the subscriber premises equipment and nearest central office or wire center. Conceptually and technically, DSL access is intended to be capable of multiple service types such as voice and data. However, implementation reality has driven most service providers to offer only Internet access without any voice service initially. It remains to be seen how long this is likely to continue. The classical telephone companies don't want to cannibalize their bread and butter—lucrative voice services—and they desperately want to tap into new revenue streams of their up and coming competitors—cable modems and DSL-capable ISPs. Therefore, initial DSL service is limited to Internet access. As the Internet matures—achieves a grade and quality of service capable of supporting voice-over IP—this situation will change. Who knows when, but someday in the future it may be possible to call up the telephone company and ask them to discontinue POTS.

Keep in mind that the main purpose of the switching function is to share use of the transmission function. Also, keep in mind the fact that change in the network is more a direct result of economic pressure than technological or regulatory forces.

CELL-BASED SWITCHING

Cell switching in communications networks can be one of two entirely different things. One use of the term applies to mobile communications backbones where the switching takes place as the mobile telephone passes from one coverage area to an adjacent area or cell. The other use of the term is more commonly known as ATM. One of the more astute and practical observations that can be made about ATM is that it's easy to get confused upfront by the name. The asynchronous part has to do with the interface between the network, which is truly synchronous and the nature of the disparate traffic, which is asynchronous, even though it maybe from a single, highly accurate digital clock source. Asynchronous also means the traffic, or data, is handled in a start-stop mode, similar to asynchronous serial interface with its start-and-stop bits.

ATM functional elements include a fixed length 53-byte cell, transmission links, and a switching machine. Unlike circuit switching, where the intelligence is resident in an underlying common channel signaling system, ATM intelligence is embedded in each cell and distributed throughout the network in edge and core switches.

Like many other technologies along the way, ATM evolved after such things as time division multiplexing (TDM), plesiochronous network hierarchy, X.25 packet networks, maybe in the same timeframe as SONET/SDH, but before wide acceptance and initial growth of Ethernet and IP networking. Early on, ATM was supposed to be the Holy Grail of all-purpose communications networks. However, that was not to be, as IEEE 802 Ethernet is now testimony to that fact, to say nothing of the past few years rampage to put IP directly on SONET/SDH transport, making IP over ATM obsolete.

ATM was crafted out of a desire to accommodate as much offered traffic as possible from the maximum possible number of users, while at the same time ensuring safe, effective (profitable) traffic movement. A casual look at the network during the time ATM was created would show a pattern much like hotels and airplanes. Equating time slots in digital transmission facilities with rooms in hotels and seats in airplanes, it was easy to see there was significant space available except during peak demand times. It could be seen that if a way to use the idle capacity could be found, it meant incremental revenue. If not, and the facilities in use could be reduced, reducing overall operating cost. Either or both of the two favorably impact the financial bottom line. So it is with Internet and Telecom facilities, not just airplanes and hotels.

From the beginning, computer communications were facilitated with either dial-up modems using PSTN voice grade services and facilities, or private leased lines. Generally the rule of thumb was to use dial-up if it was a local call. If the connection required a long distance connection, the cost was tolerated to the extent necessary to justify a full-time, private, or leased connection at a fixed cost. Even though the private line was a fixed cost and available 24/7, actual traffic passing over the facility was usually far less than full-time at whatever data rate the line was capable of.

In situations where multiple terminals connected to the same central computer, or an enterprise operated several computers in separate locations, data communications networks evolved through various forms of simple aggregate and, later, statistical multiplexing techniques. Even when a terminal is connected to a computer or system hosting an application, traffic between the terminal and host is very asymmetric. Keystroke generation is rarely more than 75 bps, compared to information and screens generated by the host application and sent back to the terminal that can require hundreds of kilobits per second, or even megabits per second. Statistical multiplexing techniques found their way into data communications equipment and networks. As more and more traffic was aggregated and transported by data networks, the same techniques were used to gain efficiencies and utilization in those networks.

Essentially, ATM was conceived and designed to cope with the bandwidth limitations in POTS and ISDN for the data communications user, while at the same time improving use of lower layer transmission facilities providing leased or private line services. The basic characteristics of ATM are built around the virtual circuit concept, including virtual paths and virtual circuits. Because source and destination information is included in each cell, switching and routing of the traffic can be accomplished by network switching equipment examining each cell as it arrives in a network, and then determining where to route it on the outbound side. This basic capability allows configuration of a virtual path through the network between any two or more points connected to the network, and to set up connections between any two or more of those locations. Thus, a virtual circuit exists inside a virtual path.

ATM also provides a capability for customer or user control of the network, enabling the user to configure switched virtual circuits within a path, and paths accommodating two or more circuits or connections. *Switched virtual circuit* (SVC) means the user pays a fee to gain access to the network, usually a fixed monthly charge based on port capacity and any local loop or access line cost. The user also pays an additional fee each time the network is used, similar to the long distance phone call model. Sometimes called *bandwidth on demand*, the service is billed according to amount of bandwidth, class of service, and time used. It can be very cost effective within a range of practical, day-to-day content transport needs.

A permanent virtual circuit (PVC) means that the carrier or service provider configures the network to provide service between two or more locations on a permanent basis. Depending on the way the service is ordered and configured, it can be flexible and complex to use, or rigid and easy to use. If the service is configured as a permanent virtual path (PVP), this means the user can configure multiple circuits aggregating up to the maximum amount of bandwidth available on the path. If that's one circuit on one path, so be it. If that's x number of circuits with equal or unequal amounts of bandwidth, that works too.

On the other hand, if the carrier configures PVC service, it can't be changed by the user, but only by the carrier after an order for changed or new service is issued. This has economic and operational implications, which may be significant or insignificant. For example, if the circuit is in use 24/7, as might be the case with a studio-to-transmitter link (STL), it may be necessary to go off air to make changes. If that is not acceptable, then establishing a new set of access and transport facilities and then moving the traffic to the new facility, followed by decommission of the other facility in previous use.

Both PVC and SVC have configuration parameters that should be considered carefully when specifying and commencing use of ATM transport. The classes of service include constant bit rate (CBR), variable bit rate, real time (VBR-rt), variable bit rate, non–real time (VBR-nrt), available bit rate, and unspecified bit rate. Each class of service has different performance and cost characteristics.

PACKET SWITCHING

One only need examine history over the past couple of hundred years to see that older communications models and methods bear resemblance to some of the current crop of fast growing methods. For example it's not difficult to see the similarity between smoke signals and telegraph messages. If one takes the smoke signal model and imagines a sender sending a message to a receiver and the receiver repeating the message to another smoke sender and so on, it's easy to see the resemblance to packet forwarding characteristics of the Internet protocol or email.

If you have any interest at all in the Internet and have done any reading on the subject, you're aware that it's based on packet switching. Packet switching depends on some basic functional elements including transmission links, and a switching engine called a router.

There are a lot of similarities between cell-based switching and packet-based switching, and there are some differences. Packet switching is simply making decisions about where to send the packet at hand. Packets are like cells in the sense that they must be opened, intelligence found about where they are headed and where they have been, and then switched and/or routed. The instructions are just inside the packet with a few other tidbits of information.

One of the fundamental differences between packet switching and cell switching is at the heart of most of the ambiguity and hand-wringing that occurs when considering the routing-switching—layer 2-layer 3 solution. It's really quite simple. Cells, PDH streams, and PPP (HDLC) are layer 2 functions. What separates these techniques from packet techniques is a time base. Packets, at least IP datagrams that make up user datagram protocol and Telecommunications protocol over Internet packets, have no reference or relationship to any clocking, timing, or basic synchronizing intelligence. They are just out there somewhere in the Ether. T1/E1, PPP/HDLC, ATM, and Ethernet all have clocking and synchronizing information embedded in the stream.

Packet switching is nothing more than switching and/or routing at packet borders, or between packets after the details in the packet header have been opened and read. Only after the entity has been opened and read can it be routed or switched to a second port. Many times the mail system is used as a metaphor for packet switching. It's a pretty good metaphor, but with some subtle differences. First, the packet entity must be opened and read. It does not have an outside and an inside unless the payload has been encrypted, or otherwise sealed and secured in some way. One of the fundamental flaws in the Internet everyone knows about and experiences every day is simple courtesy and security. In addition to the payload and addressing information, there are other significant details inside the packet entity exposed for any and everyone to see and do with as they please. These other details have to do with all kinds of fun things that can muck up the overall machinery such as administrative control of the routing machines.

A view that says the Internet has evolved from prior well-known methods and technology wouldn't be difficult to contend, but would

likely be more difficult to defend. Many modern IP network designers seem blissfully unaware that the Internet is critically dependent
on an underlying transmission infrastructure they simply refer to as
the network layer. Very few of them have a clue about the importance
of network clocking and timing. Many think packet over SONET/
SDH isn't a big deal because it's done all the time (over PDH), which
for the most part goes over SONET/SDH any way. Very few realize
that HDLC, or PPP framing, is as rigid and fixed as T1, E1, T3, or E3.
A few understand the details of packet-over SONET/SDH. The ones
that do understand this fundamental know that Internet architecture
includes layer 1 and layer 2 and is not, as the rest of their esteemed
colleagues contend, self-healing.

Throw in all the mumbo jumbo about connection-oriented and con-
nectionless protocols and mumble solution in between every fifth use
of the word router or whatever else can be thought of, but the basics
remain the same. That is, something on the premises, or at the
network access point, contains or establishes intelligence that tells
the network how to set up a connection between two or more points,
and thereby transmit and receive information through the network.
Anyone can play around with semantics all day about dumb terminals and smart networks, or at the other extreme, they call intelligent terminals and dumb networks. At the end of the day, what's
important is effective and efficient use of limited resources.

If confusion reigns, stop and ask a couple basic questions: "Is it
circuit, cell, or packet?" "What is being shared?" Is it time, bandwidth, or both? What are the interface, bitrate, and active protocols
on the facility? What is supposed to be done with it? What did the
customer ask for? What is being delivered? Is it broke? With a little
patience and perseverance, confusion will soon stop raining, the
clouds will pass, and matters will clear up as you climb up or
down the stack of bits and bytes.

HYBRID SWITCHING

Hybrid, as the name implies, is a combination of techniques. Various
combinations of hybrid switching have appeared in data communications over the years, but have been relegated for the most part to

CPE, outside the classical telephone world. In the broader context, IEEE 802.xx LLC (logical link control) and a combination of higher layer Ethernet packets mapped to Internet packets could be labeled a hybrid. However, continued innovation and development emerging after the 1984 deregulation of the long distance telephone business led to newer approaches to transmission and switching such as SONET/SDH, ATM, and other packet- or cell-based technologies. The impact of the Internet Society, in particular the Internet Engineering Task Force (IETF), on global communications standards has been significant to say the least. An example of the kind of innovation that is likely to survive and grow well in to the future is embodied in a technique called *multi-protocol label switching* (MPLS) sometimes truncated to simply label switching. MPLS is a combination of techniques derived from IP and ATM switching and protocols.

Other Emerging Switching and Routing Technologies

ASON (Automatic Switched Optical Network), GMPLS (Generalized Multi-Protocol Label Switching), and RPR (Resilient Packet Ring) are in active standards development and various stages of maturity at the time of this writing. A full description of each is far beyond the scope of this book. However, a brief description of each is noteworthy. ASON and GMPLS are often seen as a source of contention and competition between ITU and IETF standards development groups, but this is more an impression gained from reading press reports. *ASON* is a new architecture that will leverage existing SONET/SDH architecture by adding optical switching and control protocols. The motivation for ITU members and their suppliers lies mainly in reduction of capital investment at the core and operations cost associated with provisioning, while at the same time enabling paying customers to use the network in new and innovative ways. Recognize that this work is heavily biased by ITU due process in which the focus is working from well-defined detail requirements for architecture. The basic approach is to agree on the architecture and then work out a set of underlying control protocols, and don't forget that the background and experience of the developers is heavily influenced by years of operating TDM networks controlled by CCS7/SS7 and DTMF signaling.

GMPLS is an initiative that grew out of effort directed at extending MPLS traffic-engineering techniques. Recall that MPLS as a concept is something of a hybrid or a combination of circuit and packet switching. Motivation for this work lies in the simple desire to more effectively and efficiently control the transmission or transport facilities—*circuits* in the network. How to do this? Simply invent a new protocol that tells the network to "do this" or "do that" based on a new bright idea, almost without regard to practical business considerations.

The main interest driving the work of both groups appears to be some form of standardized control of the multiplexing and transmission layers in the network. Significant work remains to be accomplished before stability and maturity, sufficient to entice suppliers to invest in design and development of real products capable of attracting service provider investment is achieved. Overall, the serial cycles of standards stabilization, product design, and, finally, deployment by service providers may take 2, 3, 5, or maybe 7 or 8 years. For now, the most appropriate action for the user community is simply monitoring events as they occur.

RPR (Resilient Packet Ring) is the most mature of the three. This work is currently under the moniker IEEE 802.17. Essentially, this is an extension of BLSR (bi-directional line switched ring) technology whereby TDM/PDH, or the more common TDM/SONET/SDH, is modified to replace TDM with Ethernet, or something akin to Ethernet packets. Such an architecture would permit more efficient use of the bandwidth on a given transmission facility because it would allow multiple classes of service on a single facility. Another way of looking at it is to consider that typical BLSR configuration is such that only half of the facility is carrying live traffic, while the other is idle waiting to pickup the traffic when the load-bearing side fails. If the TDM is replaced with Ethernet or similar packet switching techniques capable of carrying classed traffic and priority based packet switching, both sides of the BLSR become useful. If one side fails, lower priority traffic gets delayed or dropped while higher priority traffic moves.

More information on these emerging standards and technologies is as close as your favorite Internet search engine.

TRANSMISSION TECHNOLOGY AND METHODS

Basically there are two types of transmission facilities: analog and digital. Transmission facilities can also be segmented according to electrical baseband, radio frequency, and optical media. In a simple third set of terms, it's either wired or wireless. (Wouldn't Marconi be proud that his term came back in vogue in the 21st Century?) Digital transmission techniques and methods are all within the OSI stack's lowest layer, most often referred to as layer 1 or the physical layer. In addition to physical transmission, or aspects related to physical movement of energy across a media, layer 1 is sometimes seen as a link in the sense that it behaves like a single link in a chain, or as a link between two other objects. In the case of transmission media in communications networks, the link characterization would apply when used to link together two switches, routers, or other network interface devices.

At the outset, telegraph and telephone technology were based on use of direct current power and metal conductors connecting transmit and receive hardware. As demand for service grew, tubes and later transistors enabled designers to place more than one conversation or transmission on a pair of wires. Thus, the communications industry became part of the electronics evolution along with radio, and later television. The telephone industry was granted rights to use parts of the radio spectrum and viewed by regulators similar to the broadcast industry as being "in the public interest, convenience, and necessity." At one time, the FCC was organized into bureaus and included separate broadcast and common carrier bureaus. Throughout the 1960s, 1970s, and well into the 1980s, microwave radio spectrum carried the majority of US long distance traffic. Radio transmission capacity became the growth engine of the long distance telephone industry. MCI applied for and received a license to carry long distance traffic between Chicago and St. Louis, eventually building a nationwide analog radio transmission network. AT&T had a combination of digital and analog radio transmission capacity when it entered into to its third consent decree with the justice department in 1982. Most of the digital radio capacity was in the access network in intra-state service. By the end of the century, those radio transmission networks have for the most part been decommissioned and replaced with fiber transmission capacity.

Until computers came into the picture, transmission was all analog. Audio, video, speech, or other information was carried in basic or baseband form on wire, or modulated on to some type of carrier—a separate, single, much higher frequency signal, transmitted, received, demodulated, and applied to a speaker or display device. Long before computers placed demands on the communications network, voice was digitized and the resulting signals multiplexed to gain more usage on common trunk facilities. As computers increased in capability and created demand for connections to terminals and other computers, the digital signal interface to the network was the natural common sense approach; however, that was not to be for many years.

Digital transmission techniques include various forms of carrier modulation in wireless, including amplitude, frequency, and phase. Modems (acronym for modulator–demodulator) are a required function and, in some cases, a separate item of equipment in satellite earth station equipment complements. The function of these devices is to impress upon, or modulate, the carrier signal with the baseband signal, or digital bitstream. Modem functions mirror, or complement, analog to digital and digital to analog converter functions. For example, an analog signal must be converted to digital before it can be transported on a digital network. A digital signal must be converted to analog before it can be transported on an analog network.

The three types of media we should know about are baseband, free space, and fiber. Most baseband transmission is on copper twisted pair or coaxial wire.

TRANSMISSION TYPE AND MEDIA

Table 4-5 presents a summary of transmission types.

Baseband Transmission

Baseband transmission is a catchall for electrical signal transport. The term baseband has deep analog roots. Two simple examples include the physical transport of sound vibrations along a string between two paper cups, and the same sound signal along two wires between a

TABLE 4-5

Media type	Medium	Application	Good characteristics	Bad characteristics
Baseband	Copper Twisted Pair	Building & Campus Network	Low Cost Easy Install	Poor Bandwidth – Distance Performance Susceptible To Crosstalk
	Copper Coax	Building & Campus Network	Greater Bandwidth Good Noise Immunity	High Install Cost Sensitive To Physical Placement
Carrier	Free Space	Inside Outside Local Long Distance	Inexpensive Path Right-of-Way	Limited Bandwidth Capacity
	Fiber	Inside Outside Local Long Distance	Huge Bandwidth Capacity	Costly Path Right-of-Way

microphone and a speaker. The first is an acoustical-based phenomena, the second an electrical-based phenomena.

Carrier transmission in simple terms is a method whereby the baseband signal is carried from one point to another on some type of carrier signal. The process of placing the baseband signal on the carrier signal is called modulation. Demodulation reverses the modulation process, separating the baseband signal from the carrier signal. Modulation techniques are beyond the scope of this book; however, Table 4-6 presents a summary of several of the most common found in use in broadcast and communications networks.

TABLE 4-6

Type	Technique	Application(s)
Analog	AM - Amplitude	Broadcast
	FM - Frequency	Broadcast
	PM - Phase	Modems
	Carrier – On - Off	Ham Radio
	PCM – Pulse Code	Telephone Service
Digital	QPSK (Quadrature Phase, Shift Keyed)	DBS
	QAM (Quadrature Amplitude)	Digital Cable TV

Free Space: Terrestrial and Satellite Wireless Transmission

Satellite transmission is just another version of, or use of, the radio or wireless spectrum. Terrestrial microwave radio and satellite radio share much of the spectrum and the equipment involved comes from a lot of common technology and many of the same sources of design, development, and manufacture. Satellite transmission is depended on very heavily in the media and entertainment industry. Almost all cable programming is distributed over satellite facilities. In recent years the direct to home or DBS infrastructure has become a viable competitor to cable and terrestrial broadcast content delivery.

Satellite transmission channels, or transponders, come in a range of bandwidths, including 18, 36, 54, and 72 MHz. The transponders can accommodate analog or digital transmission. Individual carriers, instead of a single analog signal or digital bitstream, can occupy the baseband in the transponder. The individual carriers can accommodate analog or digital signal transmission. The type of signal placed on the satellite transponder is determined by a choice of uplink and downlink earth station equipment. Most major satellite carriers have interconnection facilities to and from the global communications network infrastructure. With appropriate interface, satellite links can provide access to and egress from common carrier networks. They can also extend common carrier networks and vice versa. Satellite transmission facilities can transport standard Internet and telecom network traffic as well as terrestrial radio or fiber. But a light wave has many times the bandwidth as radio waves. A rough comparison can be found in the fact that a 36-MHz transponder with reasonable uplink and downlink antennas can accommodate a DS3 (44.736 Mbs) signal in one direction. A full duplex DS3 would require two 36-MHz transponders.

Satellite facilities are a better economic approach to point-to-multipoint service, but are only practical where there is a clear, unobstructed line of sight between ground stations and the satellite. As a rough rule of thumb, the cross-over point on something like network program distribution is around 15 to 20 multi-points where the requirement is at least 4 hours per day of usage. Once the number of locations goes beyond 20, satellite transmission is far more favorable and increases in value as the number of locations increase.

Optical transmission is somewhat different. Although there are systems that modulate a continuous light wave with a baseband signal, optical transmission techniques in contemporary communications systems and networks use a simple approach—just turn the light wave on and off at the desired repetition rate. For example, a SONET/SDH OC-3/STM4 transmit port operating at 155 Mbs turns the light on and off according to the bit pattern driving it, 155 million times a second, or thereabouts. At the receiving end, the presence or absence of light is detected, converted into a serial bit stream, and subjected to a clock and data recovery process.

Fiber Transmission

Fiber transmission is several years into its second generation, with unexploited technology in the wings to soak up demand as it grows. Fiber transmission technology and methods began appearing in the 1980s. Fiber transmission bandwidth is many times that of radio spectrum, and its bit-error performance is orders of magnitude better than any known transmission technology. Because fiber is relatively new and likely to be around in an evolving but relatively stable state for the foreseeable future, it seems prudent to explore the subject in more detail.

Basically there are two types of fiber conductors: multimode and single mode. *Mode* refers to the way light propagates through the fiber conductor. In multimode transport, the light wave travels down the fiber in multiple modes or through more than one route or path to get to the other end. In the process, the light waves traveling the different paths take slightly different amounts of time to reach the end. The behavior of the light is said to exhibit dispersion, or disperse, as it travels down the fiber. The dispersion and difference in travel time cause interference and distortion, which can impair signal integrity. Multimode fiber is the physically larger of the two, ranging from 125 to 400 microns in diameter. Losses in multimode fiber are the greater of the two, restricting its use to local applications such as LANs and building or campus environments.

Single-mode fiber is physically smaller than multimode fiber, 8 to 10 microns in diameter. This fact is the reason it's called single mode because the way to reduce dispersion and control the way the light behaves as it travels down the conductor is to reduce its size. Single mode fiber attenuation characteristics are a fraction of multimode fiber, making longer transmission distances without amplification possible.

Both types of fiber are clad with reflective material on the outside surface. This guides the light down the fiber much like a microwave radio signal travels along a round or rectangular waveguide. There are three windows in the spectrum used in light wave transmission. Short distance multimode transmission uses light in the 850-nM range. Single mode transmission runs in the range around 1310 nM and at the upper end between 1510 and 1600 nM. Early wave division multiplexing (WDM) mixed a 1310 lambda and a 1510 lambda on a common fiber in short-range applications and two widely spaced lambdas in the 1510 band such as 1538 and 1558 nM. WDM also enables bidirectional transmission on a common fiber.

An extension of WDM is dense wave division multiplexing (DWDM). Distinguishing between WDM and DWDM is more like a traveling circus or moving crap game. Claims of over 1000 lambdas on a single fiber represent one definition of DWDM, while two lambdas spaced 2 nM apart on the same fiber is another hawker's definition. ITU SG 15 wrestles with standards development in an environment where the technical landscape is still evolving at a rapid pace while business development is slow and uninteresting. How to make sense out of all this? Let's have a go at it.

First, recognize that optical fiber transmission mimics radio wave and even baseband transmission in many ways. Fundamentally the basics involving transfer of energy are applicable. Optical link performance can be modeled with similar techniques and tools as radio links. The optical transmission link has active and passive components. A light wave is generated and launched into one end of a fiber, gets attenuated on its way to the other end, and a certain level of signal is received and detected. Digital transmission on fiber is simply 1s and 0s, light on and light off. Errors in the detection of the signal at the receive end are a function of the signal in the presence of noise. If the noise is high relative to signal, then errors are more likely. If the noise is low relative to signal, then errors are less likely.

The energy spectrum of a lambda is a function of the data rate carried. For example, an OC-192 occupies more bandwidth than an OC-3, or some other data rate in between. This is the same property exhibited by radio systems when comparing bandwidth occupancy of a single DS3 compared to 3 DS3 on a radio channel, bearing in mind the fact that radio channels are intentionally and rigidly bandwidth limited to prevent out of band radiation.

Channel separation, or bandwidth between optical carriers, is as important as separation between radio carriers. If two carriers are close together, one will interfere with the other in a receiver front end and cause errors. The prevailing global fiber transmission network landscape with respect to bitrate and protocol is 10 Gbs, OC-192/STM-64 SONET/SDH on a single lambda. A transmission of 40 Gbs OC-768/STM-256 is technically mature and ready for market demand. All that speed and bandwidth is enticing and interesting to contemplate, but carriers are like broadcasters and tend not to put all their eggs in one basket. In practice, ring architecture is preferred over point-to-point links. For a reasonable incremental cost, the revenue protection provided by ring architecture is well worth the investment. As traffic demand grows, the choice is to meet the demand with an incremental physical ring, or incremental circuit cards.

OC-48/STM-16 (2.5Gbs) rings are the workhorses in the local exchange and intra-LATA networks. Given the level of traffic and revenue at stake, it makes more sense to run two or four physical rings on diverse routes, each running at 2.5 Gbs, than it does to put all the eggs in an OC-192/STM-64, 10-Gbs basket.

While it's always difficult to speculate, it seems logical to believe future growth in the transmission network will continue as it has in the past, incremental from lower to higher, favoring increments that best fit a trade-off between the actual amount of capacity needed and the risks associated with putting it all in one basket without some form of backup. If this view is valid, then it would make sense to see future use of 40 Gbs links and DWDM not as a single silver bullet solution, but as flexible tools network planners and designers can leverage to address network growth and change in existing networks, as well as preparing alternative design approaches for new networks. Perhaps supporting testimony for this view is the work that has been done and continues in the ITU on Recommendation G.692. First approved in 1998, the second Corrigendum was released in June 2002. Achieving long-term stability in standardization is challenging when the technology is not stable. And it doesn't help matters when the marketplace value is tepid or weak.

Synchronous Optical Network/Synchronous Digital Hierarchy Transmission

SONET/SDH evolved from a US industry initiative began in the mid-1980s and aimed at establishing a set of standards to support multi-vendor capability to transmit with one vendor's equipment and receive with another's over fiber media. SDH standards followed and, while they are different, the two are far more compatible than earlier T-carrier standards. SONET/SDH is truly synchronous. In fact it converts "plesiochronous" T-carrier streams to true synchronous transmission. In terms of backbone speed, it picks up where T-carrier leaves off and scales to around 40×—that's 40 Gbs.

The basic SONET/SDH transmission unit is the synchronous transport stream running at 51.84 Mbs. Most SONET/SDH transmission runs over light wave medium using fiber conductors; however, nothing prevents it from running on other baseband or carrier type medium as long as sufficient bandwidth is available.

SONET/SDH is a technology and method of assembling network equipment from diverse manufacturers to realize a connection between two points, also called a link, span, or transmission facility. The transmission facility interfaces with cross connect, cell, circuit, and packet switching facilities. With a given combination and proper configuration, SONET/SDH provides the layer 1 or physical layer connections used to support point-to-point private line, switched voice, data circuits, Internet access, or other services across town or around the globe.

SONET and SDH are very similar in their high level framework, function, and other respects. Attempting to differentiate between the two becomes an unnecessary tedious exercise for the purposes of this book. Therefore, from this point forward, the focus is on SONET and where appropriate and necessary to be specific, SONET/SDH, or SDH as a stand-alone term will be used.

SONET is an international standard; however, all international links are based on SDH standards. All SONET specifications comply with the ITU SDH recommendations. In the United States and Canada, service providers and equipment manufacturers rely on ANSI T1 (SONET) specifications. SONET standards evolved in three phases.

Phase I was issued initially in 1988 and provided eight discrete line rates between 51 Mbs and 2.4 Gbs (since extended to 40 Gbs). There were several advances over T-carrier standards, including byte-interleaved multiplexing, a monitoring scheme capable of fault alarm at path and section level, and a built-in administrative communications channel separate from the payload. Backward compatibility was ensured by creation of VT mapping to DS1/T1, DS2, and DS3/T3. In SDH, the equivalent to VT is a TU, and T3, and TU mapping to E1, E3, and E4 SDH interfaces.

Phase II enabled different manufacturer's equipment to interoperate, included automatic protection switching, and mapping of the North American DS4 (274 Mbs) signal into STS-3c.

Phase III continues to evolve in the form of improvements to past work and to address operations, administration, maintenance and provisioning (OAMP) issues, the last remaining and reportedly controversial area to be resolved.

Synchronous Optical Network/Synchronous Digital Hierarchy Layering

SONET/SDH has its own layering model. It shouldn't be confused with the OSI stack. Anything and everything that SONET/SDH does is all at layer 1 of the OSI stack. However, it is easy to see the commonality between layer 1 of the OSI and SONET infrastructure. SONET has five layers: media, section, line, path, and virtual tributary. Each of these layers can also be viewed as a level within an end-to-end facility where each level encompasses certain functional parts of the system while excluding others in the end-to-end chain. These layers and levels defined in the layering model and the end-to-end block diagram also have a distinct and direct relationship to the SONET frame, or framing structure.

First the five layers are described:

The *media layer* (also called *photonic on fiber*) is the physical media in the form of a conductor for the signal. In practice, the signal traveling over the conductor is either radio or light wave energy.

The *section layer* carries STS frame and section overhead information such as error monitoring and scrambling.

Line layer contains the synchronous payload envelope (SPE) and its overhead. The SPE basic channel is a 51.840-Mbs continuous bitstream. This stream takes on a OC or STS label. For example, a single stream is called an OC-1 if it is an optical signal, or STS-1 if it is an electrical signal. This basic channel can be multiplexed with others running at the same rate into higher order streams. When this happens, it takes on a number indicating the number of basic streams, such as OC-3 or STS-3, running at 155.520 Mbs.

The *STS layer* lies between the line layer and the virtual tributary layer. This is the demarcation between legacy T-carrier (T1, E1, E3, DS3) signals, functions, and services, and SONET functions and services.

VT layer is the interface between SONET and legacy T-carrier network facilities.

A block diagram of a multi-section SONET/SDH layer is shown in Figure 4-6. The level of detail includes the basic components of a typical link. Each of these components are defined and described in terms of their relationship to each other and what they contribute to overall link function and performance.

SONET link components include the media, repeaters, light wave terminating equipment (LTE), and terminal multiplexer (TM).

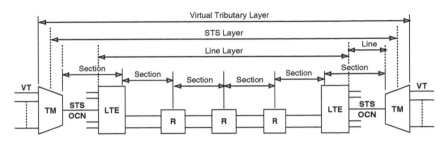

FIGURE 4-6   Basic SONET Link Block Diagram

Components in the media layer include fiber cable with multiple fiber conductors, regenerators, light wave amplifiers, conduit, poles, and other supporting structure. If a facility uses regenerators, it will not include light wave amplifiers and vice versa.

The line layer includes all the components in the media layer, plus light wave terminal equipment at each end of the facility. The function of these units is to convert electrical signals to light wave signals. They also act as an add/drop multiplex unit whereby one or more STS can be broken out of the backbone and used to originate and terminate traffic from both sides. Another way to view the LTE is to say that they accumulate and map STS signals and multiplex them into a backbone signal. On the other side, or in the form of a reverse function, they break out STS signals and pass them to terminal multiplexing TM or add/drop connections in the LTE.

The terminal multiplexing equipment builds up and breaks out STS signals. Another way of saying this is that the TM aggregates traffic from legacy digital sources such as those that originate in voice grade circuit switching and cross-connect equipment. It also accommodates STS-1 or 51.8 Mbs streams as a peer to legacy T1, E1, E3, and DS3 traffic streams. Other non-PDH layer 2 traffic can also be accommodated such as ATM, asynchronous, unchannelized, E3, DS3, and packet-over SONET or point-to-point protocol on high level data link control (PPP/HDLC).

There are other important functions involving monitoring, alarms, network performance, and network management, including support for OAMP.

Media layer components consist of fiber conductors, regenerators, light wave amplifiers, conduit, poles, and other supporting and protection infrastructure objects.

SUMMARY

In summary, transmission systems have evolved over time beginning with basic or baseband electrical transmission. As electronics evolved, radio or wireless transmission evolved and served communications growth very well. Just when it seemed like the radio spectrum was becoming congested and in need of capacity, satellite technology provided much needed relief and new ways of serving the broadcast industry as well as competition for cable. Currently, media, entertainment, and communications industries are agape as the public takes to the Internet like a duck to water. During the 10-year close-out of the 20th Century, fiber transmission techniques and methods matured and continued to make transmission capacity available to meet demand for what seems to be an endless appetite for movement of information and access to knowledge.

Around the time of the break up of AT&T in 1984, the public network in the United States was essentially a flat network offering voice and data services over PDH transmission. Voice services were provided by almost 100% analog switches in the end offices, interconnected to the long distance network. Competition in the long distance network was just beginning to emerge. Much of AT&T's long distance network was digital, using PDH T-carrier techniques. The smaller long distance start-ups used stored program controlled analog switching machines.

Data services were almost exclusively facilitated with private line-services over bandwidths ranging between 2.4 and 56 Kbs. In rare situations, 56-Kbs data traffic would be aggregated and carried on T1 backbone facilities. Figure 4-7 shows a high-level topology for the US Public Communications or common carrier network around the time of the break-up and divestiture of AT&T, and the beginning of communications deregulation in the United States.

During the course of significantly less than 20 years, the communications landscape changed dramatically. PDH transmission, limited

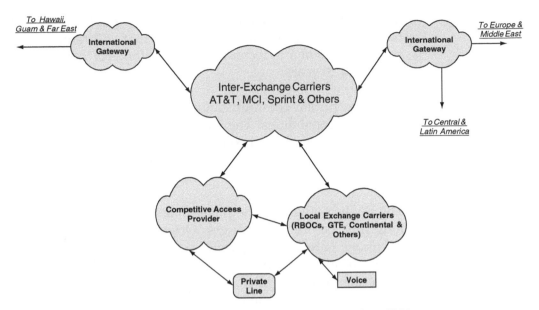

FIGURE 4-7 US Public Communications Network, c. 1984

to approximately 1 Gbs using proprietary techniques, would be replaced by standardized byte-interleaved, synchronous transmission running at 40 Gbs. Voice traffic was being rapidly overtaken by data traffic. Something called the Internet was 20-plus years out of a Department of Defense initiative, the objective of which was to create a survivable network in the event of catastrophic damage to a single site housing network components passing inter-nodal traffic.

By the turn of the 20th century, it seemed like everything using wired transmission was going wireless and everything using wireless was going wired. Standards development that took three to five years was eliminated or brought to reality in drastically reduced time frames driven by special interest groups and new players. Rigid, government-regulated, tariff-based services crumbled in the face of new classes and groups of service providers built around incumbent service providers and their new competitors. Traditional PDH facilities, in particular private leased line services, became the growth engine for the classical telephone companies, many of which were, and still are, prohibited from offering anything other than dial tone and relatively low-bitrate voice grade services. New services at layer 2, 3, and 4 of the OSI stack became "overlay" networks designed, built, and operated by organizations of all kinds. Figure 4-8 shows how public network topology has changed. This is the infrastructure that will remain in place and continue to evolve as technology and methods of using it evolve in the next 10 to 20 years.

The new public communications network includes more capacity and more capabilities. The shortcomings of TDM, or T-carrier, and circuit switching have been overcome with SONET/SDH transmission, cell, and packet switching technology, except where analog or PDH runs alone in its own environment.

The public network is the physical layer for the Internet. It will be a long time before that changes. Originally, the physical layer in the telephone network was passive wire conductors, later bundled into cable sheaths, and over time, evolving to other types of media, including radio and fiber. The architecture of the network's physical layer can be separated and classed into access and transport

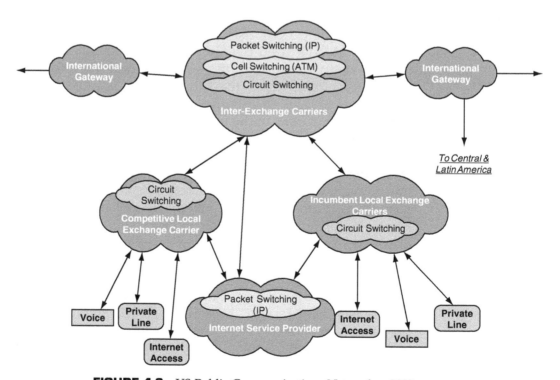

FIGURE 4-8 US Public Communications Network c. 2003

facilities. The philosophy behind this architecture arises from the processes by which services are designed, used, and charged for. All end-to-end service originates via an access facility and terminates over a transport facility at the other end via an access facility.

There are two types of facilities: switched and private lines (Table 4-7).

Switched access is limited to low-bandwidth telephone service and Internet access provided by ILEC, CLEC, and cable operators. Classical voice grade dial service using dual-tone multifrequency (DTMF) signaling supports connections that enable voice conversations and Internet access—"computer-to-computer" communications up to 56 Kbs. Higher speed services are available from local exchange carriers and cable operators using DSL and cable modem technology, respectively. DSL is capable of a few hundred kilobits per second, while cable modems achieve speeds well over a megabit per second.

Private line access is required to use any form of higher speed switched services from all service providers. All the service providers use equipment from one or more manufacturers that make

TABLE 4-7

Facility or technology	Facility type	Frame duration	Frame type	Packet size	Bandwidth	Data rate
PDH/SONET/SDH	Transmission	125usec	Fixed Timeslot	fixed	64Kbs-10Gbs	fixed
Cross Connect	Circuit Switch	125usec	Fixed Timeslot		64Kbs-10Gbs	
POTS	Circuit Switch	125usec	Fixed Timeslot	fixed	64 Kbs-1.5 Mbs	fixed
ISDN	Circuit Switch	125usec				
Frame Relay	Packet Switch		frame	variable		variable
ATM CBR	Cell Switch	125usec	cell	fixed		fixed
ATM VBRrt	Cell Switch	125usec	cell	fixed		variable
ATM VBRnrt	Cell Switch	125usec	cell or PDU	variable		variable
ATM ABR	Cell Switch	125usec				
ATM UBR	Cell Switch	125usec				
Ethernet	Packet Switch		frame	variable		variable
IP Routing	Packet Switch					
MPLS	Packet					

standardized subscriber interface at layer 1 and 2. Layer 1 will be T1/E1, T3/E3, or at the lower SONET/SDH OC3/STM1. Layer 2 will depend on the capabilities of the subscriber and the service provider's equipment at each end of the access facility.

A variety of technology is employed in access and transport facilities that enables all the various services available in any given service provider's network. The service set offered depends on the configuration and physical proximity of the service provider's switching and transmission systems. Table 4-7 summarizes the major technologies used in switching and transmission facilities and shows framing, packet, bandwidth, and data rate characteristics. View this summary and the table as an introduction to more details about facilities and services covered in the next chapter.

Broadcast Industry Services

Broadcast services are something of a special case. From the early days of radio, AT&T built transmission facilities to support broadcast networks. Splitters in the form of resistive networks or transformers, amplifiers, and equalizers enabled multipoint connections from the network source to all the radio stations making up the network. By the time television networks required distribution facilities, AT&T's network contained significant broadband capacity in the form of frequency division multiplexing over microwave radio and coax transmission backbone. Carrying television audio and video became a matter of interfacing the frequency division hierarchy in much of the same conceptual ways it's done today with digital networks. Satellite transmission, with its lower cost and reliability advantages, quickly caught on and replaced AT&T terrestrial services for national network distribution.

Services in a local area used similar equipment, but the transmission media for audio was a standard copper twisted pair wire like that used in telephone service. Television program transmission was mainly on microwave radio or baseband transmission on 124-Ohm balanced coaxial cable.

With the advent of digital networks, so-called encoding and decoding equipment was designed to convert analog program signals into digital signals and interface them to the digital network. Typical radio program material requires several hundred kilobits of bandwidth depending on the nature of the signals and number of program channels. Similar equipment carries NTSC or PAL program material on an E3 or T3 transmission channel.

References

1. OSI/IEC 10731, http://www.OSI.org/OSI/en/OSIOnline.frontpage
2. http://www.bipm.orglen/scientific/tai/time_server.htm/
3. "Fundamentals of DS3." Telecommunications Techniques Corporation Whitepaper Application Note, 1992; found on: http://www.acterna.com/united_states/technical_resources/downloads/white_ papers/fund_ds3.pdf

5

NETWORK ARCHITECTURE, FACILITIES, AND SERVICES

If network architecture can be thought of as analogous to a framework, then facilities and services can be thought of as the bits and pieces that give the architecture detailed substance and a useful purpose. In communications networks, as in broadcast networks, there is an input and output—the basic function. Both must be managed and their assets and cost of operation must be accounted for. Both are enabled with interfaces and protocols. Both tend to evolve and adapt to regulatory and technology forces. More importantly, those that respond favorably to market demand, survive, grow, and prosper.

This chapter is not about how to design and build a telecom network. The intent is to provide information about the nature and characteristics of networks sufficient to communicate desires and requirements to potential equipment and service providers using Internet and Telecom terms, standards, and symbols. The overall goal is to help the reader understand communications network architecture, facilities, and services. The secondary objective is to provide a framework with which to evaluate potential service provider's ability to deliver facilities and services capable of meeting their requirements, and ultimately assessing and measuring performance against the terms in a contract.

LANGUAGE AND TERMINOLOGY OF COMMUNICATIONS NETWORKS

Communications network architecture is most often taught and thought of as being layered and linked. Layering breaks the network into a set of logically related components. Linking connects the components and makes up an end-to-end facility or service. The network becomes the link between all the equipment at all the locations making up the network. Successful use of network facilities and services always includes a set of common equipment at each location where facilities are terminated, enabling service delivery. Still, equipment, facilities, and services alone are insufficient to ensure successful business use. Man cannot live by bread alone; he must have peanut butter. The key ingredient required for success is a walking, breathing human being, qualified to be responsible and held accountable for making sure all the equipment, facilities, and services are configured, maintained, and changed as day-to-day demands of the enterprise change.

If you are a designer, manager, or senior staff responsible and/or accountable for making or approving decisions related to design, planning, and management of assets and operating expenses related to communications network asset and expense management, the material in this section is one of the more important, if not the most important, part in the book. It can be a technical foundation for a process whereby you and your organization can take firm control of communications equipment and service vendors by first defining your needs in their terms and lingo so you can establish a competitive procurement environment as the first step in building a long-term mutually beneficial relationship between your organization and the vendors. You can benefit greatly if you take the time to study and understand the technical and economic aspects of how to order and piece together end-to-end facilities and services to build the network best suited to your business.

The level of detail in this chapter is structured around the common equipment at each location and its interface to network facilities and services. One side of the equipment is connected to the network facility that provides access to network services. The other side of the equipment interfaces to local area network (LAN), private branch exchange (PBX), and Moving Picture Experts Group (MPEG) compression and decompression equipment.

Figure 5-1 shows an example of the common equipment found at a typical operations site engaged in production, post-production, and on-air transmission operations.

Basic service requirements include telephone or dial-up service, data transmission, Internet access, and content transport within and outside each site. Group ownership operations may range from half a dozen operational sites to 50 or more. The geographic scope might range from purely local stations to statewide or regional or national in size.

This level of detail may seem insufficient when considered from a broadcast operations point of view; however, it is quite satisfactory for network planning and design purposes where the objective is to inform a potential equipment and/or service provider. It can be used to delineate equipment from services, and it can also be used to inform potential consultants or network designers about their responsibilities in terms of a design, project, or program management.

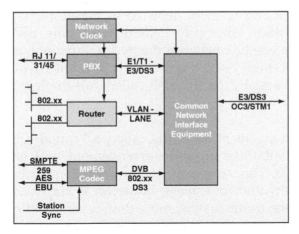

FIGURE 5-1 Typical Operations Site Common Network Equipment

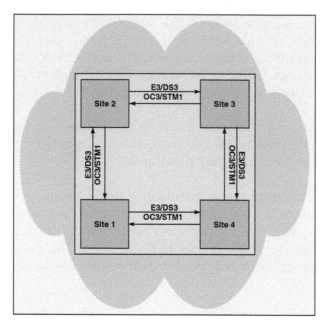

FIGURE 5-2   A Four-Site Network Topology Diagram

On the communications network side, the level of detail goes substantially deeper in the form of definition, description and use of various network elements, and several alternatives for linking them together. Figure 5-2 is a simple sketch showing four sites with links between each.

The link arrangement depicted in Figure 5-2 is only one of several possibilities. Obviously if there were only two sites, there would only be one link between the two. Traffic requirements would drive the selection, but it's quite likely the links would be bidirectional, and of equal capacity, also called full-duplex, asymmetrical bandwidth.

The arrangement shown is called a "round-robin" because it connects all the sites in a series arrangement, with only one link between any two. This arrangement has operational advantages in terms of reliability and robustness; however, it may have disadvantages because the traffic may be more than it can handle in some places and more than needed in others. For now, keep in mind that there are three types of traffic: voice, data, and program content.

While not explicitly mentioned, the Internet is becoming much more important in many ways; however, it will be included as a potential resource when the traffic is segmented into content creation, distribution, and delivery. Also, don't forget that the Internet is nothing more than just another resource built on a set of technology. The technology is the family of Internet protocols (IP) that can be used separate and apart from the public Internet to build a private network based on IP standards and techniques.

Regardless of the basic technologies and all the architectural considerations, it is the end-to-end service between any two sites that we seek. Figure 5-3 is simply a reminder of our reference model.

Any seemingly complex, multisite network can be decomposed or decoupled into single, defined point-to-point paths, which can be observed, monitored, and measured in many ways. Once end-to-end measurements are made and a profile of its characteristics recorded, it becomes a benchmark for future operations.

One last point is that the single path can be broken apart and individual component performance measured and characterized. One simple obvious characterization is connecting two sets of equipment with wire or fiber, sometimes referred to as the "perfect network," and measure performance. And of course there are ways to measure and characterize the performance of the facilities and service making up the link between the sites.

Now we go to details of network architecture, facilities, and services.

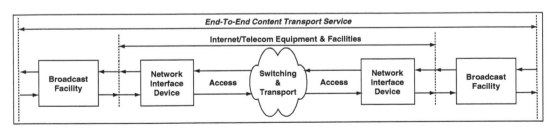

FIGURE 5-3 End-to-End Service Reference Model

NETWORK ARCHITECTURE DESCRIPTION

Communications network architecture is generally thought of in terms of access, interface, geographic coverage, media type, and signaling or control systems. The basic functional parts of the network architecture are premises, equipment, access, and transport facilities. Access and transport facilities are made up of routing, switching, transmission, clocking, and network management elements. Today's network as well as the network of the foreseeable future will continue to carry voice, data, video, and the Internet. Today's network includes the Internet and the global public and private telecommunications network it is built on. Within that network are facilities based on wire (coaxial and fiber), terrestrial and satellite wireless technology, and methods.

LAYERING AS USED IN COMPUTERS AND COMMUNICATIONS NETWORKS

The basic idea behind layering is that computer equipment and system functions are bound by, or within, layers. Each layer is bound to, or interacts with, its neighbor immediately above and below. If each layer up and down the stack interacts, or interoperates, with its neighbor successfully, then the system or process making up the overall system or network is likely to succeed in performing all the functions it was designed to accomplish.

The layering concept was created by the International Standards Organization to serve as a standard definition of computer industry structure dealing with communications issues in computer environments. The standard is named the open systems interconnect (OSI) model and typically referred to as the OSI stack. The communications industry adopted the technique for use in standards and design documents. Two examples of communications stacks are shown in Figure 5-4 along with the OSI stack. The OSI stack has seven layers; the synchronous optical network/synchronous digital hierarchy (SONET/SDH) and Internet stacks are both four layers.

Taken in isolated display, there doesn't appear to be much of a match or direct relationship between the three. However, we know that

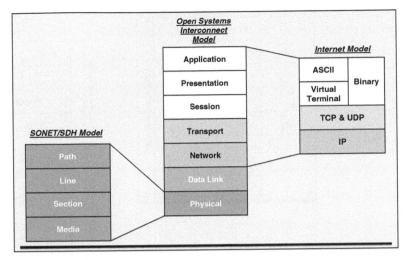

FIGURE 5-4   Layering Models Used in Computing and Communications

SONET/SDH is an example of the physical layer of the OSI stack, and only the physical layer.

The four-layer Internet stack used by many rests on an IP layer, which is a peer to the network layer in the OSI stack.

Each of the layers in the OSI stack is unique in function and behavior, and has a specific relationship with its neighbors above and below. The OSI model is subdivided with layer 1 and 2 classified as being hardware-oriented, whereas the upper layers are said to be software-oriented. As an entity, the lower layers are communications-oriented; the upper layers are user- or application-oriented.

Another commonly used way of illustrating the layering concept and relating it directly to linking is to draw an end-to-end service model with multiple layers showing multiple virtual channels on top of a physical channel. Figure 5-5 is another example of the use of layering models to show relationships between the layers in communications networks and computers.

Figure 5-5 has been redrawn from the three separate models shown in Figure 5-4. It shows these three models and how they relate to each other.

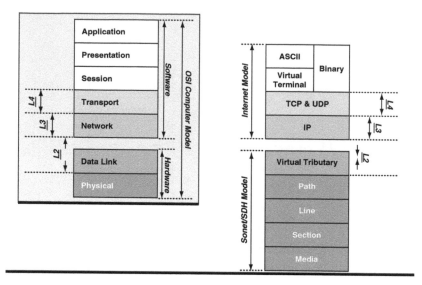

FIGURE 5-5 Example of Three Layering Models

The OSI model has been separated between layer 2 and layer 3. This is the division between hardware on the lower level and software on the higher level. The Internet model is left intact and placed above the SONET model. The VT layer has been added to the SONET model.

Layers 2, 3, and 4 have been shaded with four different levels or shades of gray, reflecting the dividing layer between hardware and software in the OSI model. There is a similar division between the IP, and lowest layer in the Internet model and the VT/STS, or highest layer, in the SONET/SDH model. The data link layer is a point of commonality across all three. That is, voice and data networks use exactly the same physical layer network interfaces and protocols as the Internet does. For example, an unchannelized E1/T1, E3/DS3, or OC3/STM1 facility can be used to support all types of traffic defined in the site location architecture.

Most depictions of SONET/SDH layering don't include the virtual tributary (VT) layer. If this layer is included on top of the SONET/ SDH model, it becomes a direct fit, or interface to the Internet model for plesiochronous (PDH) point-to-point links or leased/private line facilities. Another point worth mentioning is the fact that many purveyors of network equipment and services are offering direct interface between IP and SONET/SDH synchronous transport streams. In and of itself, it's an incremental step. However, if this physical layer capability is combined with differentiated services in the network, traffic aggregation, and type of service (TOS) capabilities in new and emerging network equipment, the result is a potentially dramatic and profound impact on ISDN/PSTN and plesiochronous (PDH) network facilities and services.

Consolidating mixed or disparate traffic requires knowledge of all three layers and a detailed understanding of how they relate to each other when integrating equipment and software. More importantly, understanding how the traffic payloads are organized and structured is key to successfully mapping the traffic to the network to get maximum use of the network. Much has been said and written about convergence or converged networks. Occasionally, the term *multi-service network* is used. Layer 2 and 3 is the place in the layering models where multiple, disparate traffic types are converged and mapped to a common access and/or transport facility. Physical placement of routers and switching equipment and its configuration determines physically where, and in what sequence, the so-called convergence (of disparate traffic) takes place.

Applying the layering concept to communications networks seems quite natural and logical. Two types of layering are commonly used to depict communications networks. Classical telephone networks are structured around a multiplexing, switching, and transport hierarchy, while computer networking and the Internet are structured around protocols and interfaces.

Another term that creeps into the lingo from time to time is *overlay network*. For example, the larger multi-service carrier networks share transmission facilities between voice grade services that require 64 Kbs transmission links with ATM switches and IP routers that use raw transmission bandwidth in varying amounts. Since the ATM and IP networks came into existence after voice grade services, they were built and are said to overlay voice grade services. Occasionally this term is also understood to mean the ATM and IP networks are above voice grade services in the OSI stack, where voice grade services are seen at layer 2, ATM at layer 2/3, and IP at layer 3/4.

Continuing the evolution, the Internet community has taken these two separate models and created a separate but related structure dubbed *Inter-networks*. The Internet is a completely different structure with its own unique behavior in terms of how it moves a payload, otherwise called *content* or *information*. The concept is built around an idea that combines payload information with address information, hands it to the network, and the network not only carries the information, but when it gets to its destination, the network communicates that fact back to the sender. Overall result: A third network is built using the same kinds of standard layer 1 and layer 2 facilities as voice and data networks.

Continuing to add communications facilities for separate network applications has led many to question and wonder if there may not be a better way to organize their traffic to get better use of all the facilities being paid for. Attempts to address these issues have led to solutions called *converged* or *multi-service* networks. A more appropriate label might start with the traffic whereby disparate traffic types are converged into a common access facility or across a common transport facility. Viewing the three layered models in a single context can be a constructive and instructive step to defining requirements for a network capable of carrying disparate traffic.

LINKING IN COMMUNICATIONS NETWORKS

Linking involves two types of paths through the network: physical and virtual. Paths are created through the network when physical links are connected in series between two terminal sites. A receive path output from equipment in a terminal is connected to transmit path input on another facility. For a full-duplex or two-way path, the receive path from the opposite direction must be connected to the transmit path in the opposite direction. This type of linking process in digital networks can be extended many times without undue service impairment, except for the accumulation of transmission delay and errors. The latter can be mitigated by good link engineering practice common in radio and optical link budgeting.

When any network is made up of three or more sites, another term comes in to play called *meshing*. Networks are either fully meshed or partially meshed. The four-site network depicted in Figure 5-5 is classified as partially meshed because there is no direct path or link between sites 1 and 3 or between sites 2 and 4. Figure 5-6 shows a fully meshed four-site network.

The difference in meshing has implications in terms of economics, reliability, and robustness. Economically, the partially meshed network connected in a round-robin fashion requires four links. A fully meshed network would require six links. The two additional links might increase the monthly charges for leased private line facilities by 50% when comparing a fully meshed, four-site network to a partially meshed network. Such a choice or decision is the eternal dilemma of network architects and designers. The way out of the woods requires economic analysis and judgment to resolve. Chapters 11 and 12 provide a process and tools for resolving technical dilemmas with financial approaches. For now, simply note that two-site networks need not be concerned about meshing issues. Three-site networks require four links to be fully meshed, four-site networks require six links to fully mesh, and a five-site network requires 10 links. Figure 5-7 shows a five-site fully meshed network.

As the number of sites in a network increase, the number of links soars, dramatically increasing complexity. The more complex network

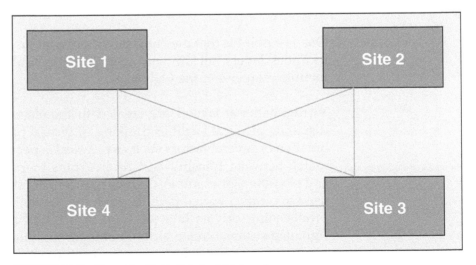

FIGURE 5-6 Four-Site Fully Meshed Network

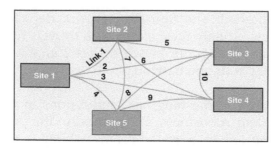

FIGURE 5-7   Five-Site Fully Meshed Network

architecture becomes, the greater the need for more detailed documentation required to manage cost, reliability, robustness, and network performance. Another factor mitigating the need for fully meshed networks is use. For example, consider that when one site connects to another, it is for the purpose of passing traffic. That traffic may be flow only one way or both ways, and if the equipment or people at the site are busy or engaged with one site, it may not be possible to connect to a third site simultaneously. From a slightly different perspective, if two sites are busy, then the likelihood of either of them becoming engaged with a third site is limited. Therefore, use of the other possible links is significantly less likely, so why spend money on all that capacity. However, without a fully meshed network, ways to communicate between all the sites must be established, and the answer is switching. The carrier might provide the switching function, or it can be done with premises equipment.

One last point is that meshing might occur at any one of the communications layers, all the way from the physical layer through and including layer 4 of the OSI stack.

Virtual paths at layer 1 are created in the electrical domain, across disparate physical facilities. Linking of virtual paths involves a combination of several factors such as physical aspects of the connector, a match between transmit and receive pins or pairs, signal polarity, and clocking. For example, digital cross connect switching systems are used to create a virtual path through the transmission network when provisioning private line services. POTS/ISDN switches and their signaling systems create virtual paths through the same transmission network to support voice grade connections enabling telephone calls, facsimile, and modem transmission. The mechanics of the connection involve connecting a transmit/receive pair on one side of the switching system to a receive/transmit pair on another facility.

ATM network architecture includes a virtual path layer, inside which virtual channels or circuits are created and placed. Routing and switching are performed according to information in the virtual path identifier and virtual circuit identifier sections of the ATM cell header.

The linking term is also applied to a process or protocol to create a path for data or information between disparate media. A link created with IEEE 802.2 logical link control (LLC) at layer 2 is at the highest layer of LAN architecture. It defines a set of protocols that support services between the media access layer and the transport layer. LLC is functionally equivalent to the telephone hook-switch and DTMF keypad used to control setup and teardown of a voice connection or link. Most LAN cards, and many other devices supporting LAN connections, have a green indicator light. If the light is illuminated, it indicates physical connectivity between two devices. Many telephones, especially those with two or more lines, have an indicator to show off-hook, active line, or in use. Most equipment with a wide area physical connection is equipped with some kind of indicator as well as alarm to indicate status of the link.

CLOCKING CONSIDERATIONS

Clocking and timing in the public network is based on four levels of accuracy, traceable to worldwide time standards. These highly accurate clocking sources are generically referred to as stratum 1 through stratum 4, where stratum 1 is the most accurate. Before 1984, the US portion of the network was primarily the responsibility of AT&T. Network clocking information was distributed throughout the network infrastructure whereby a single clock signal source was sent through the network hierarchy. The most accurate clock signal was distributed to a set of switches that passed it on to lower levels in the hierarchy with the lowest and least accurate clock signal reaching end office switches serving subscribers.

This all changed in 1984 when the network was broken up into the seven regions and equal access to the long distance network was declared. The seven regions were divided into 200 plus local access and transport areas (LATAs) where traffic originating and terminating in different LATAs had to be transported by the long distance carriers. By the time of the breakup, the other long distance carriers already had networks running from independent clocks. With AT&T's network running on its own clock, this left each of the seven local exchange carriers no choice except to design and build their own clock systems into each LATA's network.

Original T-carrier or T-1 clocking was a simple matter of two channel banks inter-operating with each other over a four-wire connection. As long as clock stability and recovery were within range, there was no problem. When the clock signal ranged out of bounds of clock recovery, something called a *slip*, meaning an out-of-lock condition occurred, followed by re-synchronization of the receiving clock. From a practical perspective, this causes a barely audible click in speech. The impact is hardly noticeable in telephone conversation and will likely cause a re-send in data applications, but it can be a killer in content transport applications, especially where the application is something like an MPEG transport stream.

Over time as digital components made their way into switching systems and higher-order multiplex equipment, clocking signals were distributed through the network itself. When that happened, all of a sudden, network clocking was dependent on the network. If the network failed, then clocking failed. To solve this issue, the concept of clock hold-over was created. Hold-over simply set some design and operational rules in place that said, "When clocking signal is lost, switch to a local source and hold the stability within specified range until the master clock signal is restored, then switch back to the master clock signal."

Most pieces of equipment in use have at least one clock. Typically, the design includes external reference capability and configuration parameters that allow the user to control the operation so that it either stays on internal reference, or alternatively locks to an external source. If it is programmed to lock to an external source and it fails, the equipment automatically reverts to the internal clock source.

Communications network clocking architecture is multi-level or strata. There are four levels: stratum 1, 2, 3, or 4. Stratum 1 is the most accurate and stable. Stratum 1 clocks, by definition, do not rely on and may not have or need external references but simply an ability to be calibrated on some periodic basis—say on 1-year intervals. Stratum 2, 3, and 4 are less accurate and stable than stratum 1, and they typically have provisions for external reference input.

From a strict timing perspective, stratum 1 clocks are at the core of the network, while stratum 4 clocks are at the edge of the network such as in a PBX, channel bank, data multiplex equipment or router. Going back through the chain, stratum 4 clocks depend on a stratum 3 clock, which depends on a stratum 2 clock, which depends on a stratum 1 clock, the master clock for the network.

It's a given that long-term, undisturbed connections through a network depend on continuing, long-term, undisturbed common clock signals and their distribution (or linking to every single network element), which carry a digitized payload. Numerous strategies are used or chosen from to formulate clocking in a network. Regardless, the network operates synchronous or non-synchronous. T-carrier networks are said to run plesiochronous. Taken literally, and in subjective practical terms, it means nearly synchronous. Therefore, it is not synchronous, as understood by an experienced radio or television engineer.

Such operation is characterized by the fact that it experiences clock slip behavior sooner or later. Clock slips mean that the clock and everything that uses it vary from its reference by an amount sufficient to cause the payload to be out of coincidence with its original clock by one clock cycle. It's not a matter of whether or not clock slips occur, but when. Even when all the network elements run with external reference to a common, higher order accuracy and stability clock, slips can happen from various effects such as temperature changes and atmospheric variations in the case of wireless media. Truly synchronous transmission media became a reality with SONET/SDH network transmission standards and along with that came a significant improvement in network stability.

How to deal with clock distribution and stability is an overall, end-to-end network issue. Dealing with it from a design level perspective requires obtaining a set of facts about the capabilities and limitations of the common network equipment and the network service provider's clocking and distribution infrastructure, and the specific part of that infrastructure providing use of facilities and services to all the sites included in the overall network. More specifically, decisions about network interface equipment should not be made without a clear definition and understanding of the performance characteristics of the equipment with and without external clock reference. The second most important piece or set of information is a definition and understanding of the communications network-timing source and its relationship to the network channel interface at each site.

Once a picture as outlined above has been obtained, it can be assessed in light of the requirements for content transport. The broad perspective includes a situation on the one hand where it may be acceptable to rely totally on service provider network timing sources as the master timing and clocking reference, while the opposite extreme is to equip each site with a set of stratum 1 clock references. Obviously, somewhere in between is likely the practical approach for the enterprise. Don't forget that network clocking and synchronization is an entirely different matter than station clocking and synchronization. As long as the communications network runs on an accurate enough clock and assuming it is configured to carry live or real-time program content, the content will be clocked from the originating site, with its embedded clock intact, into the communications network, and then moving it from there to one or more sites, extracting it from the telecom clock time base and transferring it to the time base of the receiving site.

One last tip is the fact that a communications network referenced to a global positioning source (GPS) or other source of universal time becomes attractive if the content being transported is also GPS referenced. A good middle of the road, practical design trade off is to reference the MPEG codec-to-station sync and reference the network interface equipment to the service provider facility. If the network bit-error rate is of sufficient level, then the network will be stable and always timed to the same source as the station is. If clock reference is lost, then the network becomes subject to the internal references of the interface equipment, and clock slips become more of a problem because they cause disturbances in picture and sound quality and may cause loss or corruption of closed caption or other similar content.

COMMUNICATIONS NETWORK SWITCHING AND TRANSMISSION ARCHITECTURE

Telecom network architecture in 2004 is very much a child of many years of growth and technical evolution driven by telephone calls. Telephone calls require 64 Kbs and are point-to-point connections used for short periods. Telephone service economics where the longer the distance between two points of a call costs more money influenced most of the traffic into local and regional calling patterns. These regional calling patterns led to hierarchical network architecture with high capacity switch sites in fewer numbers piecing together circuits to enable regional and national telephone calls.

Figure 5-8 is an illustration of the hierarchical nature of the telephone network.

The access tandems and certain designated transmission facilities are generally referred to as the *access network*. These switches formed the technical and economic basis for a philosophy the FCC dubbed *equal access*. Next time you pay a telephone bill, look at the term FCC access charge. Nope, the money doesn't go to the FCC. It goes into the owner of the access facilities. And according to FCC regulations, the owner of the access facilities in each LATA must afford equal access to any inter-exchange carrier (IXC) willing to arrange for a connection between their facilities and the local exchange carrier's (LEC) facilities.

On the other side of that equation, equal access means the LEC must offer their subscribers the ability to select their long distance, or IXC.

The higher levels in the hierarchy vary, depending on the size and extent of the individual long distance carriers. AT&T operates much as it did before divestiture. Others, such as MCI and Sprint, have flatter or fewer levels in their switching hierarchy. All the long distance carriers have gateways to international networks and their own shares in international transmission facilities. Piecing together two

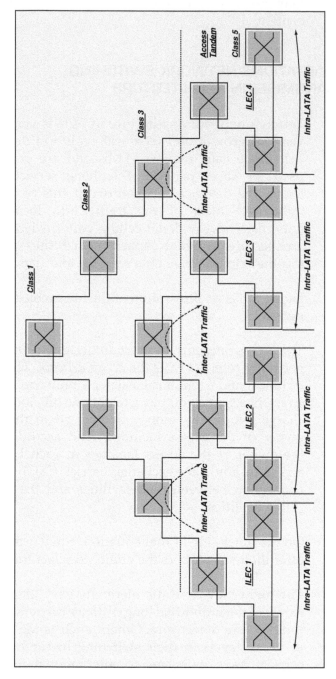

FIGURE 5-8　Classical Telecom Hierarchical Architecture

or more SONET links end to end creates an extension of the physical boundaries of the network with the addition of more links (Figure 5-9).

All the layers are extended across the interface. Each layer becomes an extension of itself each time a SONET transmission facility is connected to another. Linking two or more in series along a path or route, extends the boundary or domain as defined in the facility specification. Figure 5-9 shows a block diagram of SONET transmission infrastructure linked to provide continuous bandwidth that could be used to provision an E-1/T-1/J1 or E3/DS3/J3 private line. Such a facility could be used as a link in a data communications network or to connect two routers in an inter-network or intra-network link.

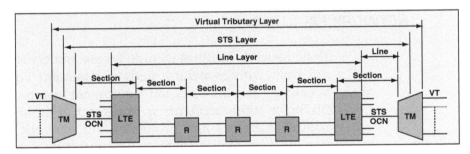

FIGURE 5-9 SONET Transmission Segment

Another method of linking facilities is the mid-span meet, commonly used by carriers to link or interface between networks. This is depicted in Figure 5-10 and shows links between LECs and IXCs.

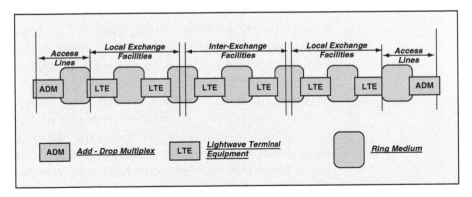

FIGURE 5-10 Linked SONET Transmission Facility

A variation on the light wave terminal equipment (LTE) is the add-drop multiplexer (ADM). SONET transmission capacity is built around the 51 Mbs STS. An LTE breaks out all the STS in the optical carrier (OC) and makes them available. If the LTE is used at a location where all STS are not terminated in local routing/switching equipment, then any transit streams must be connected directly to other LTE facing other routes. The ADM breaks out and terminates a limited quantity of STS from the optical carrier. For example, the office in which the terminal is located might only require 1 or 2 STS from an OC-3, OC-12, or higher rate transmission facility. Using an ADM would be less costly in terms of capital investment and operations expense.

NETWORK FACILITIES

Basic network facilities include access and transport. These facilities are built with switching and transmission equipment. Generally speaking, the transmission equipment meets certain ANSI, ITU, NEBS, or other standards with respect to input and output signals, control interfaces and protocols, and network management. Since deregulation and divestiture, manufacturers have sought market acceptance of products across a combination of public carrier networks and private or enterprise networks. Corporate buyers not in the network services business have continued to press for improved network operations and uptime. These basic trends have driven manufacturers to design and build market specific products from building block designs.

Complementing these trends, carriers, and service providers are adopting and adapting products to meet demands for service and facilities beyond traditional network boundaries. Many carriers offer managed services where the service provider designs a network, procures, installs, and then manages equipment and services on a life-cycle basis. This changing environment has led to approaches that include a mix of equipment and services designed to meet operational requirements of the business or organizational enterprise. Many state governments have undertaken or plan to undertake outsourcing initiatives.

CARRIER NETWORK OR ENTERPRISE NETWORK?

Obviously if you're in the broadcast business, it's not likely you're a common carrier or, maybe stretching the point a bit further, you're not in the Internet service provider business either. However, that said, it doesn't mean you can't build the same type of network used by these types of business. Quite often the term carrier class network, or enterprises class network will be seen or used. Both terms are mostly marketing hype; however, there are ways to distinguish between them. It's important to decide first of all what your requirements are. In fact, in some content transport applications, neither class of network is satisfactory, and networks built to specific performance levels are required. Call it what you will—it's still a network, and it had better support or meet the business's needs or else the business may not be a business very long.

FACILITY TYPES

Three types of facilities may be used to make up a network. The simplest network is built using point-to-point private line facilities and premises based multiplex and switching equipment. The most complex networks are made from a combination of point-to-point private lines, premises based multiplex, and carrier provided switching equipment. Essentially there are three types of facilities: access, switching, and transmission.

Access facilities are used to connect the customer to the carrier's network at one or more locations. Access facilities are also commonly called *local loops*. Because of the fact that these access facilities serve only a single customer, they are typically a dedicated private line between the customer and the nearest central office. Access facilities are usually provided by a local exchange carrier and almost always connect the customer to a long distance carrier, or IXC. Charges for access facilities are based on a single one-time installation charge and a fixed monthly charge.

Switching facilities are made available by either local or long distance service providers from a set of switching systems shared among many customers. Switch facilities can be programmed or configured to switch traffic within a private network connecting multiple sites of a single customer, and provide access to long distance service outside the network. Switching facilities typically include a port or connection charge, plus a usage charge each time the user accesses the network and uses it for whatever service has been arranged for.

Transmission facilities are the pipes used to connect switching facilities, classically referred to as trunks. You may hear the term *access trunk*, which refers to the customer side of a switch, or *inter-machine trunk*, which refers to a connection between two switching facilities. Charges for transmission facilities are similar to access facilities, a one time installation charge, followed by a fixed monthly charge based on bandwidth.

NETWORK SERVICES

Network services are established after equipment and facilities are installed and connected together. Network services require the equipment and facilities to be configured to meet specific service description parameters. One of the most critical tasks in designing and building networks is preparation of a service description for each service required. Too often, this step is ceded to the service provider or inadequately documented by either the customer or the service provider, and the results are less than satisfactory in terms of the resulting business impact.

NETWORK CAPACITY

Capacity is a measure of traffic or payload carrying capabilities. *Throughput* is sometimes used as a measure of the ability of a circuit, facility, or network service to carry bits. Different traffic types have different constraints based on their nature. Continuous, live, or pre-recorded content requires continuous, uninterrupted channel bandwidth equal to or greater than the bandwidth occupied by the content. Telephone calls, audio, and videoconference materials are of the same nature as live or pre-recorded content. Discontinuous signals such as hypertext markup language pages and file transfers operate just fine over continuous bandwidth; however, continuous bandwidth may impose economic penalties compared to discontinuous bandwidth facilities.

NETWORK PERFORMANCE

Network performance metrics include availability, reliability, robustness, and bit error rate. Obviously, these terms are subjective as single words. Including them in a contract for service is wise. Defining them in words that can be used to set performance expectations where money is to be paid in exchange is even wiser. Drawing a diagram and making a list of locations and facility or circuit identification numbers that the definitions cover is absolutely necessary to ensure services received meet expectations; otherwise, money may be paid for inferior or unsatisfactory services.

Here are some generalized statements in layman language to serve as a starting point for definitions in a service contract.

Availability is a measure of the time network resources being paid for that are actually available for use. The time period may be as specified as 24 hours a day, 7 days a week, 365 days a year.

Reliability is a measure of the time the network is performing in accordance with specifications when it is available.

Robustness is a measure of the network's ability to recover from loss of service caused by a malfunction of one or more elements making up the network.

Bit error rate, also referred to as *bit error ratio*, is a measure of bit errors received, compared to the total number of bits received over a given time. Packet networks measure lost packets instead of measuring bit errors. The packet either makes it through the network, or it doesn't. Obviously, if the network doesn't successfully carry a packet, there will be errors in a file.

Transmission facilities in communications networks are built on terrestrial and satellite radio wave (wireless), wire, and light wave media.

In general, wire has significant distance limitations depending on the electrical bandwidth and level of signal it's expected to carry. Radio waves are better than wire in both respects, but not nearly as capable as light waves. Light waves have many times the bandwidth capacity as radio waves.

For example, broadcast satellite service (BSS) facilities are designed around an architecture having 32 transponder channels, each with 32MHz occupied bandwidth. Each channel carries a single quadrature phase shift-keyed bitstream capable of about 30 Mbs payload. Direct satellite service (DSS) facilities occupying a single slot are typically configured with 36-, 54-, or 72-MHz transponders and are typically loaded with subcarriers. Each subcarrier is separated from its neighbors, depending on the amount of bandwidth in the payload it's required to carry. Generally, a 36MHz transponder can accommodate a DS-3 (44.736 Mbs), depending on uplink antenna size, transmit power and receive antenna size. These variables all accumulate in a calculation involving reliability, robustness and bit error rate (BER). The lower the error requirement, the more expensive the facility becomes. Terrestrial radio operating in the same band, for example, 4 to 6 GHz or 11 to 12 GHz exhibit similar payload capacity and BER characteristics. Radio channel baseband BER ranges from 10^{-7} to 10^{-9}.

Wireless LANs are proliferating as well, using the 1Ghz spectrum. Bandwidth ranges from 10 to approximately 50 Mbs, depending on equipment and physical free space path distance.

Single mode fibers running SONET/SDH optical carrier–based protocol has many times the payload capacity and significantly better BER. An OC-48/STM-16 runs at 2.5Gbs and has a capacity of 48 STS-1 (51.84 Mbs). Each STS-1 can carry 1 DS-3 and several more T1s. Baseband link performance is in the range of 10^{-9} to 10^{-11} BER.

When differentiating between satellite and terrestrial transmission, it's important to consider several factors. These include the nature or topology of the network, reliability, robustness, and error rate performance. All things being equal, which of course they aren't in every detail, but at a high level, satellite networks tend to be more cost-effective and reliable in point-to-multiple-point topology than terrestrial networks. This is not a fiber vs. radio technology consideration; it's a physical consideration. Fundamentally, the satellite transponder is the common source for multiple receivers of the same information. The physical aspect is one transmitter (albeit an expensive one) to many inexpensive receivers. If the satellite could radiate light waves instead of, or in addition to, radio waves, the consideration wouldn't change. What would change is the payload capacity for a given radiated source. It's well known that light waves are capable of carrying far more information than radio waves. The difference is primarily the fact that free space propagation or attenuation is drastically different than physical conductor propagation or attenuation.

Network architecture is best described with a topology map or block diagram. Depending on the business requirements and nature of the traffic, the topology map and block diagram can be combined in a single figure and be used as a conceptual tool or have greater detail added and be used as a design detail document in test and acceptance work, network performance evaluation, or fault diagnosis. The topology map or block diagram is used to capture and convey locations, number of sites, type of facilities, number and type of connections, service demarcation points, bandwidth, phone numbers, network domains, IP address, service provider boundaries and interface points, type and class of service, etc. Figure 5-2 is an example of a topology map or diagram that could be used to describe a network's topology and services required to support content transport, voice, data, and Internet traffic.

6

DEFINING NETWORK APPLICATIONS

Application definition is perhaps one of the most critical work activities or efforts in the practice of dealing with Internet and Telecom assets and resources. Define the application properly and it can significantly improve enterprise operations and profitability. Define it loosely, incompletely, or improperly and it could stifle growth, stop the business, or prevent a new venture from getting off the ground.

Defining communications applications requires business and technical knowledge. It's also helpful to have experience defining applications in a particular field. For example, defining Internet and telecom applications specific to the manufacture of widgets may work well in a business manufacturing fishing line, sinkers, and tackle boxes, but someone with this background would be in for significant study before successfully designing and building a network for a regional bagel bakery. However, if the regional bagel bakery wanted to expand to a foreign country and required a network, someone with experience building a network for the foreign expansion for a widget or fishing gear manufacturing enterprise would be a good candidate to undertake the work. Building global or international networks across government boundaries requires specialized knowledge of disparate regulations as well as communications service provider capabilities and practices across these same boundaries and within the host countries of the enterprise sites involved. The process of defining communications applications is fairly simple if it is started with a clear understanding of the enterprise operation contemplating investment in equipment and expenditures for services. There are two basic sets of conditions. Understanding and documenting these at the start may seem trivial and useless, but spending the time to get it right up front won't take long and may prevent a business failure in the future. One set of circumstances involves a complete new, clean sheet of paper design and implementation, such as starting a new business. The other, in its simplest form, is an incremental expansion of an existing business or operation. The basic difference between the two is rooted in the fact that a start-up will have far greater one-time expenses and overhead than an incremental expansion. It's like adding a room to a house, which requires wiring and outlets that can be supported by adding one or more circuit breakers to an existing circuit breaker panel connected to an existing electrical service entrance facility, capable of supporting the new requirement for power.

WHO ARE YOU AND WHERE ARE YOU COMING FROM?

There's an old homily that says that nothing ever got finished until after it got started. Not only is it important to understand and document your situation, it's equally important to understand your capabilities. Wanting to do something or having a curiosity, fascination, or interest in a bright idea is a far cry from being handed a well-defined business need by a manager or executive in an organization. The other side of that coin is getting a brainstorm or brilliant idea, presenting it to the manager or executive, and then being told, "Okay, go do it." However, recognize the potential trap in the words *well-defined*. Who's the subject matter expert?

RESPONSIBILITY AND ACCOUNTABILITY

The story is told of a wise and kind queen seeking a replacement for a retiring coach driver. Over a period of several days she interviewed many candidates, arriving at a short list of three who were truly outstanding and really interested in the position. Another set of interviews with each candidate in private, re-affirming their experience, skills, and knowledge of roads, trails, bandit gangs, and sources of food and water throughout her land and neighboring kingdoms only made the decision more difficult. So the Queen decided to call all three in for a side-by-side comparison, hoping there would be some differentiating characteristic to make her choice less difficult, and let the candidates see the same so as to ease the task of telling the losers they didn't get the job. There were three outstanding candidates, all physically about the same size, handsome, quiet, confident—perfect examples of humanity. On first appearance, her dilemma seemed to be worsened. Then, as happens to all good queens when needed most, a bright idea appeared. She decided to tell all three about one place along a regularly traveled road over the mountains, around a curve, and near a sharp cliff several hundred feet above the valley floor. Then she asked each to say how close to the edge of the cliff they could safely drive the coach. One volunteered to answer first and said, "I'm a very skilled driver, and I can drive your coach within a foot of the edge of the cliff and you'll remain safe." The first candidate's answer encouraged the second to speak up and tell the Queen not only was he a skilled driver, but he was kind to horses and made sure they were always fed, watered, and taken care of before thinking of himself. Possessed of a competitive spirit and wanting to differentiate himself from the other outspoken candidate, he confidently told the Queen he was sure he and the team could drive the coach within 6 inches of the cliff—better than 50% closer and get her to her destination safely! Recognizing a big difference and starting to feel comfortable that a clear choice was emerging, she turned to the third candidate for his response. Looking the Queen squarely in the eyes, he said, "Your highness, I'm a skilled and careful driver as well. And I'm not intimidated by risk; however, I tend to avoid it as a general practice. When the team and I are responsible for your safe travel on that road, we'll stay as far as possible from the edge of the cliff." Guess who got the job?

The moral of this story is to illustrate the point that, as the subject matter expert, you are responsible and accountable. The Queen in our little story didn't know what she didn't know. But it wasn't difficult for her to learn. It's highly likely she wouldn't have learned, except one out of three candidates was knowledgeable and willing to speak from experience. The message is to be careful because you are accountable and may pay with your own posterior extremity, when the Queen loses hers while traveling the fast lane a little too close to the edge.

So what's your personal role? Are you the manager or executive whose role is to approve a bright idea, or are you handing a bright idea to someone else to implement? Are you the subject matter expert—the knowledgeable and experienced coach driver taking an opportunity to clarify and polish a bright idea into a clearly defined statement of work?

DEFINING BUSINESS REQUIREMENTS

Business generally refers to a commercial or industrial establishment. Business also refers to the activities of an organization. Great differences exist between General Motors, the Catholic Church, and the US Federal Government. For sure, their missions are not the same, but they all share a need for communications networks with somewhat different requirements. In recent years, *corporate network* is out; *enterprise network* is in. *Communications* is now the domain of advertising, marketing, public relations, investor relations, and employee relations. Telephone service and data networks have morphed into Telecom. The Internet arrived on the scene and is now, to the great delight of its inventors, disrupting everything.

Business communications requirements should address or respond to the needs of the business. Every organization, large, small, or in between, has a need to communicate internally among its people, often called "our most valuable asset," and externally with current and prospective customers and suppliers.

Communications networks by whatever name have been influenced by computer software and lingo. Lots of computer programming is done using the problem-solution model, wherein no significant programming activity is undertaken until a clear, definitive problem statement has been created. This influence or perhaps unfortunate adoption of a practice has lead to unfortunate situations where solutions are bought and paid for before it's discovered that no problem exists. Or, worse yet, a solution was bought to solve a real or perceived problem that was ill-defined, and the real problem didn't go away. In severe cases, real problems can kill an organization before the learning curve catches up and a fix is found and put in place. Partial fixes can be deadly as well. Buying a DS-3 network interface card without an attendant upgrade of a T-1 to a DS-3 access facility doesn't relieve any of the pain of congestion for Internet users.

If problem-solving enables successful computer programming, successful network and systems design relies on a well-defined statement of requirements. When real business needs drive communications requirements, most problems are anticipated and avoided. The problem of having paid money for an ill-defined or undefined solution is avoided. Successful computer software and communications network planning, design, and building start with definitions that ultimately enable predictable outcomes. Define what you want up front. If you have been asked to do something and given a requirement, you must read, understand, and think through how you would define the requirement. If you can see ways to improve the requirement, suggest or incorporate them into the requirement. If you are asked to undertake a task or implement a project, and a requirements document was not provided, document your understanding of the requirement, hand it back to the requestor, and make sure that's what they want before setting forth to build or acquire. Lastly, it's not productive to extend this process. Sooner or later a point is reached where this step is complete and ready for the real work.

Highly summarized, all organizations require communications capabilities beyond simple face-to-face and written communications. In modern communications network terms, that's phone service, computer communications, and, in the broadcast industry, capability to transport content during content creation, distribution, and delivery operations.

BUSINESS REQUIREMENTS TEMPLATE

A business requirements statement includes introductory information about the organization such as official corporate name, address, and contact information of the representative who will speak, act, or obtain approval to act or otherwise commit the organization to accept and arrange payment for goods and services. This introduction should also say something about the nature and mission of the organization. A 24/7 operation responsible for public safety, convenience, and necessity has different requirements in terms of reliability and robustness than an 8 to 5, 5-day-a-week operation. While this kind of general information may not be specific, it rounds out the document and orients buyer and seller thinking.

FINANCIAL CONSIDERATIONS

Laying out broad financial requirements establishes an awareness of the importance of managing operating cost and capital investment. Every organization has unique financial characteristics. Leasing equipment instead of buying may be preferred, or it could easily be the opposite, depending on the state of the organizations finances. Non-profit organizations are always open to donations of equipment and services and depending on the tax status of the supplier, it may be opportune for manufacturers and service providers to exchange goods and services for tax receipts. Details such as invoicing for payment of taxes, requirements for insurance, and a myriad of other terms and conditions included in standard purchase orders should be introduced as appropriate.

TECHNICAL REQUIREMENTS

Depending on organization and project size, design resources, budgets, and facilities management practices, natural divisions of responsibility may occur between facilities management, communications, and computer work groups and require close collaboration and cooperation. Defining communications applications is a matter of considering the technical aspects and characteristics of equipment, facilities, and services needed to satisfy the business requirements.

Like the layering models, it's useful to picture communications network applications at the lower levels, and business or organizational functions at the higher levels. Each location where communications network facilities and services will be used will require physical layer network access and transport. Physical layer access may only appear in one single room in the establishment, but higher layer functions are likely to be spread around the entire building or campus with appearances in each office, cube, conference room, operations center, equipment room, or even outside in parking lots or production areas.

Once an understanding of a business need is clear and mapped to equipment, facilities, and services, it's helpful and enlightening to analyze the traffic by breaking it out by type across organizational and/or operational function. Voice, data, and Internet access are likely to be at all workstations and many items of shared equipment; however, it's unlikely that anyone except key individuals in payroll, accounting, and human resources would have access to payroll information. Core content transport may be limited to certain operational functions and even compartmentalized to limit individual access in certain cases. It is also helpful to characterize each application with a priority. For example, all the traffic from all the users necessary to get a commercial on the air would logically take priority over downloading a new commercial from an ad agency website.

The selection and invoking of standards helps to define applications and facilitates acceptance. For example, once a supplier has been qualified, inspection and verification of compliance can be conducted on a periodic sample basis instead of dealing with each and every telephone instrument, network interface card, or other high quantity items. Circuit and facility acceptance test and verification can be reduced from 72 to 48 or 36 hours. Simply stating that compliance with a particular standard, engineering guideline, or recommended practice can serve as a shortcut or alternative to multiple pages of detailed specifications.

If the requirement is for additional facilities at an existing service location, circuit and facility inventory records should be studied for potential use of spare or unused capacity. New circuits and facilities should be planned with the idea that it may be possible to migrate and consolidate all traffic on new facilities, allowing decommission of part or all of older facilities, and reducing operating expense.

DOCUMENTING TECHNICAL REQUIREMENTS

Specific technical requirements should be accumulated in about three or four documents, depending on size and complexity of the project, including a project outline and description, network topology, spreadsheet or matrix containing cost and technical parameters, equipment, facilities, circuits, and floor space plans.

All documents should be created as living documents available on one or more server sites and in paper form. Depending on the maturity of the project, changes, revisions, and releases may vary. Change is a fact of life and must be coped with at any time. Early in the life of a project, changes are likely to occur hourly, daily, or weekly. Once in operation, changes will be limited such as when people move to new work locations, or an equipment update is brought on line.

Organization of the spreadsheet is probably the most critical. On large projects—tens of sites, hundreds of workstations, terminals, pieces of equipment with network connections—it may be wise to use a database manager instead of a spreadsheet, or begin with a spreadsheet and convert to a database manager. Database managers are awkward to work with in the early stages of planning; spread-sheets are awkward to work with as reporting tools as a project grows and after it goes operational.

Whether using a spreadsheet or database manager, a matrix or table style document is the core of all design documents. Organized correctly, it will support directory services, operations analysis and reporting, and many uses not thought of when it was created. The spreadsheet should be organized with location, department, operations function, and users on the vertical axis and equipment, service, and applications on the horizontal axis. Application is meant in the broadest possible way and should be subdivided by specific name, such as under telephone, include class of service, extension, terminal number, and other details deemed important.

Terminals and workstations should have similar information up to around OSI model layer 3. This should dovetail and not substitute for other information above that layer included in IT/MIS documentation. It's likely some duplication of information will occur, simply to make it easier to get at for maintenance, but organizational ownership and design authority should be clear.

Sheets containing cost information and other unit detail can be a valuable tool and support high-level summary sheets. Once a spreadsheet grows beyond a few hundred lines and more columns than easily fits on a page, and/or more than half a dozen summary sheets, it's time to think database manager. Don't forget to keep multiple copies in separate physical locations, under controlled release.

CAUSES OF CHANGE IN EQUIPMENT FACILITIES AND SERVICES

Dealing with changes in communications is a never-ending, challenge. A work group outgrows its physical space and new space must be found. A new employee is hired and must have workspace. The reverse is also unfortunately true and requires action by the communications manager. New technology constantly makes opportunities to improve the way business is done.

Coping with these changes in the media industry presents new and unique opportunities. Content creation, distribution, and delivery, once wholly analog and transported with physical or real-time transmission media, are well along the path to a complete, end-to-end digital environment. Internet technology, products, and services have added new dimensions and challenges such as asset management and security. Commercial transactions where transfer of content and rights to use it are exchanged for money on an individual or one-at-a-time basis make the business look like telephone calls (i.e. keeping track of small, valuable transactions that when aggregated, billed, and collected, become a significant business).

Networks designed and used in everyday business to support voice, data, and Internet access are not satisfactory for content transport because of the amount of bandwidth required, reliability, robustness and network performance in terms of bit error rate. However, once a network is designed to deal with content transport, voice and data represent incremental levels of traffic across a more reliable and robust network at lower unit cost.

Depending on the nature of the content, coping with requirements can range from situations requiring an immediate need for breaking news to well known and scheduled events such as the Olympics, political conventions, and elections. Elections represent a special challenge. Much is known, or strongly suspected, about the locations where the news will originate, but the sheer magnitude of the number of locations coupled with the unknowns about the outcomes make it easy to miscalculate availability of circuits and facilities, wreaking havoc with competitive news coverage.

COMMUNICATIONS REQUIREMENTS FOR PRACTICAL BROADCAST OPERATIONS

Communications requirements in the broadcast industry include voice, data, Internet access, control functions, and content transport. It's helpful to consider the business or organizational functions requirements separate from content creation, distribution, and delivery operations. Ongoing, day-to-day production of content by the networks (ABC, BBC, CBS, NBC, PBS, HBO, etc.), and local stations, broadcast radio, TV, and cable, have unique requirements for telephone and data services to "get it on the air." Special events are just another way of doing the same basic work in a different location. Getting the job done, or getting the program on the air or in the can, requires at least two—sometimes three—people for simple, one-personality stand-up coverage. Capturing the content or recording, it alleviates the need for live transmission. Voice communications may be facilitated with two-way radio, or mobile telephones.

VOICE SERVICES

Live, on-air events require voice lines for communications between producer, program director, camera operators, technical support, and coordination between them and their counterparts at the remote site. Depending on the complexity and number of people involved, several private line voice grade circuits may be required.

DATA SERVICES

Any number of possible production operations may require so-called data services. For example, the FCC requires control and monitoring of transmitter equipment. Character and graphics generator systems have to be programmed, and this may require access to computer systems, other data storage, or the Internet. Videotape recorders, electromechanical beasts as they are, require commands to play, record, rewind, or fast forward. More often than not, these commands come from editing control or automation systems.

INTERNET ACCESS

Internet access for media industry businesses, broadcasters especially, is a challenge for several reasons. First, assuming they want to survive, the Internet is a new transmission medium for their programs and advertising, capable of doing all they do and more. If they don't defend their position, their competitors will get their audience and advertising dollars will follow. All of a sudden the local newspaper has audio and video in addition to still images and print. The radio station can do images and print. Television stations can now make their radar weather displays available 24/7, not just during the regularly scheduled weather report or emergencies. So-called streaming video is a matter of how many programs, when, and available or obtainable capacity.

Internet access can be used to support internal operations—moving content via file transfer. And, of course, anyone with a computer needs access to be able to cruise—get access to information and applications on local servers, corporate servers, or outside the firewall. Lastly, who can do without email?

CONTENT TRANSPORT

Where does your organization's content come from, and where does it get delivered? If you are a typical broadcaster, it is likely you are involved in some kind of multiple-station operation such as a group ownership, local marketing, or operations agreement. Moving content around on network facilities provides the ultimate in speed and flexibility. Defining the amount of content and the timeframe required to support existing program schedules is a first step. Once the level and timing of traffic is defined, then basic decisions about how to implement a network can be considered. Fundamentally, the approach will likely involve incrementing existing equipment, facilities, and services. In any case, the common-sense approach is to fall back on the basic models defined earlier in the book. Once the technical definition and traffic parameters are known, then cost and revenue impact assessments can be made. If and when a decision to go with a project is made, then the next step involves pricing, availability, spending money, and taking risks, all of which are explained in Chapters 8, 9, and 10.

7

SPECIFYING EQUIPMENT AND SERVICES

Specifying equipment and services can make or break a project, even an established organization, if not properly structured and carried through the procurement process with professionalism and care. Too many buyers don't specify properly and wonder why the suppliers respond in kind. Many buyers go on a fishing expedition with a broad shopping list they perceive to be a request for proposal (RFP) when at best it is only a request for information (RFI). Perfectly intelligent but woefully unexperienced and lacking knowledge, individuals pass themselves off as consultants to managers and executives who are ill-equipped to pass judgment on their ability to create and document requirements, specifications, conduct evaluations, and make recommendations throughout various aspects of a procurement process. Worse yet, many organizations buy and pay significantly more than they should or would have to if they were to take the time to prepare and carry out a reasonable procurement process. The process of procuring goods and services is only one aspect of a project or program. Likewise, specifying equipment and services is only one aspect of the procurement process. This chapter is about specifying what is to be acquired. The focus is getting the specs right, one of the most critical steps in any successful project or program. If what you want is not properly specified—the specs can sour an otherwise flawless acquisition process. Acquisition of goods and services is the subject of the next chapter.

CONSULTANTS

Every profession has its quacks and incompetents. Internet, Telecom, and computer industries have their share. When times are good, competent consultants are busy. When times are bad, competent consultants are busy. It's not difficult to engage an incompetent consultant in good times or bad. Like other professionals, engaging a new consultant is similar to finding new medical or dental professional after a move.

Specifying and procuring professional labor services can and should be treated like any other service. If your organization engages accountants and attorneys, you can easily engage a communications consultant. Most third-party professionals are engaged through direct knowledge or by referral. In other words, if you don't know an individual who can help get the work done, ask someone you know and trust if they know someone. Finding and hiring or engaging a consultant starts with the initial contact by whatever means, and from there on it's like hiring a new employee.

If you are not sure how to define what you want, that's a perfectly good reason to hire a knowledgeable consultant. If you know what you want, and you just need someone who can take direction and get the work done, that's another perfectly good reason to hire a hardworking, knowledgeable consultant.

Like any procurement, the ground rules are the same. Specify what you want in writing, hand the document to the potential consultant, and get them to confirm their understanding of what is written. Then agree on an hourly rate and an estimate of time and dollars for each of the tasks of your project. Consultants can be hired to perform single or multiple tasks. The key is to be specific about the task they are hired to perform. If you're uncertain about what you want done, or how to go about it, competent consultants can help you define it and then you can hire the same consultant as a follow-up engagement or bid it out. Consultants don't like this sort of approach for obvious reasons, but from a purely hard-nosed business perspective, it's a matter of being satisfied that your organization's money is well spent.

One of the most common uses of consultants is to prepare specifications. This works well if your organization has formal procurement rules and policies. If not, then the next most common use of consultants is preparation of a RFP. The work of the consultant can end at this point in the overall process or extend to evaluation of responses to the RFP, contract negotiations, and supplier management.

Expect components of your project to vary. Simple projects are more like tasks. As cost, scope, and complexity increase, multiple tasks turn into a project. Sometimes multiple projects are dealt with as a program. For example, ordering and turning up a T1 or DS3 private line connection between two sites can easily be seen or labeled as a task or project that can be completed in a few days or weeks. On the other hand, converting a collection of point-to-point private lines to frame relay or Internet access based connectivity between a data center and three remote sites is more like a project because of the level of capital and operating cost, amount of time, and resource level required to accomplish it. Broaden the imagination slightly and visualize converting an enterprise network from classic circuit-switched voice services and frame relay connectivity to an IP network with differentiated services capable of transporting voice, data, content, and providing fire-walled access to the Internet, and it begins to sound like a technology conversion program, especially if private branch exchange equipment is replaced with session initiation protocol telephones connected to local area networks (LANs).

If your organization has embarked on a major project or program, engaging a qualified consultant can be appealing, not only in terms of getting the work done, but in introducing change in the organization. For example, it's quite common for management to be convinced that investment in new technologies and capabilities is valuable (i.e., there's a return on the investment). But optimistic action turns to skepticism and outright pessimism at the mention of increased headcount to execute and operate. Well-qualified consultants can fill the gap and become a powerful way to educate, inform, and change attitudes, which is the first step in changing the way business is done.

DEFINING TASKS, PROJECTS, AND PROGRAMS

Defining work to be done is another way of defining a *deliverable* in contractual lingo. A deliverable can be a completed work item as easily as it can be a piece of equipment or software placed on the receiving dock. Accepting a deliverable as complete and satisfactory becomes the basis on which a transaction is moved from money-due to money paid status. Ordering a single network interface card, or a gross, is easily done with a single purchase order. As long as they are all being delivered to a single address, installing and configuring them is a labor intensive deliverable. If a third party does installation and configuration, it's likely the value is many times the cost of the network interface cards, yet all three items can be labeled a deliverable in a purchase order and paid for separately on acceptance.

Defining a project is more complex because it involves multiple tasks and deliverables. Projects may be segmented into phases built around time frames and capabilities. The data network conversion from private lines to frame relay mentioned previously might be further segmented into phases timed across several weeks or months while the technology conversion program might be structured in phases built around voice, data, and Internet access capabilities.

Structuring a project or program, like defining an application, should logically start with a statement of requirements. The statement can be used in a proposal in response to a verbal request, or it can be posed in a written document by the requestor. If the statement of requirements is structured properly, it can flow smoothly through a response to a request (a proposal) to the negotiation and acceptance of the proposal, which becomes the basis, or in fact, a contract, between two parties. The work proceeds through deliverable, acceptance, payment, and contract completion. Contracts for goods are usually considered complete on delivery and acceptance, except for any warranties covering defects in workmanship and materials, or clauses covering patents, copyrights, or other intellectual property.

The same process is used for equipment and services. The difference between the two is mostly a matter of physical items or rights to use physical items such as software or intellectual property where the fee paid is for rights to use instead of ownership. Services are tangible, yet no physical property is included in the transaction. A specification for equipment and services is a matter of defining precisely what and how many or how much is to be included in the deliverable.

Depending on organizational policy and practice, standard terms and conditions of purchase on the buy side may conflict with standard terms and conditions of proposals on the sell side. It's even more likely that standard terms and conditions of a buyer will not be in complete alignment with standard terms and conditions of the seller's proposal and/or sale. When software is the deliverable, license or rights to use the software can become very complex.

FORMS AND INSTRUMENTS USED IN EQUIPMENT AND SERVICES PROCUREMENT ACTIONS

Several types of documents are likely to be used in equipment and services transactions. Some of the more common include RFI, RFP, request for bid (RFB), request for quotation (RFQ), and responses to each of these: proposal acceptance form, a purchase order (PO), standard terms and conditions of sale, standard terms and conditions of purchase, invoice, receipt(s) for deliverables, payment, check, irrevocable letter of credit (ILOC), promissory note, and others. Equipment and services specifications are likely to be included in, or referred to, in all of these.

RFB, RFI, RFP, and RFQ are all terms having much in common with terms such as network, quality, and others where their meaning can vary drastically, depending on who's using them and what their purpose is. It's worth a few lines to set out how they are used here, and establish what is meant by each.

RFB means that what is being bid on is well known and quantified in terms of deliverables and qualifications of potential suppliers. It's most often used by municipal, state, and federal government entities where the practice is to use standards and detailed technical specifications to define materials and processes instead of using commercial off-the-shelf product names. Constructions of a water tank, courthouse, or a highway are examples of the outcome of a project covered in a request for bid. Another name for an RFB is an invitation to bid (ITB). Bidders are invited to bid based on their presence on a list of qualified bidders. Maintaining a presence on such a list is based on past performances and admittance to a list is by a due diligence process. An ITB or RFB usually implies all approvals and funding is complete and minimal or no development or technological issues are known. Credible responses are in strict point-to-point form, and each paragraph or point in the response must be addressed and only with the information asked for. Unsolicited or irrelevant information may be used to disqualify the respondent from consideration.

RFI means the requestor is not sure what they want, nor are they committed to any action whatsoever. They are quite likely to simply take the information, examine it, and do nothing. On the other hand, they may take the information and use it as a basis for a large complex project or program requiring senior management or even board approval.

RFP is perhaps the most common of all similar actions and something of an "in-between" to the extent that the issuer knows generally what it wants and intends to budget and or buy. Final approval to commit may be subject to senior management or board approval. It means the buyer wants and invites creativity in responding to the request form and contents. RFP actions provide the most negotiating flexibility for buyer and seller.

RFQ means "What is the price and when can you deliver or ship?" It is used most often in transactions involving commodities or in cases where the requirements are specified in detail sufficient to quote firm pricing and shipping information.

SPECIFICATIONS

Given all the procurement and contract forms and instruments mentioned previously, specifications are usually referred to as a document. A schedule of equipment and services to be supplied is another example of an important document that must dovetail closely with the specifications document. In journalistic terms, the specification defines what, and the schedule defines how many of each. As the name implies, the purpose of the specification document is to be specific about equipment or service functions, capabilities, and performance. Well-written specification documents contain an introduction, a description of how the equipment or service will be used, and how it must perform within the specified limits to be acceptable. It is very important to be specific and realistic. Realistic, from the perspective of business need, means what the business can tolerate across a range of performance, and realistic about what can be potentially delivered. Making the performance tolerance too tight, or difficult to achieve causes cost and delivery time to increase. Making it too loose lowers cost and time to deliver, but it might create performance results that don't satisfy business need and risks quality and customer acceptance on the output side of the business.

EXAMPLE EQUIPMENT AND SERVICE SPECIFICATIONS

The remainder of this chapter contains examples of specifications that can be used to detail technical requirements for a digital transmission facility for a multiple station group. The examples cover a studio-to-transmitter (STL) network, digital television transmitter, design, erection, and installation services. These should be read and used within the context of the overall block diagram referenced at the beginning of Chapter 3. The material is intended to serve as the technical content of an RFI or RFP and acceptance of deliverables covered in Chapters 8 and 10.

STL Network

The STL network will provide and all transmission facilities required to transport content, monitor and control transmitters, and support voice and data communications, site security, and network management functions. The network must be designed to meet very high reliability operations with less than a 1-minute outage per month, cumulative outage of less than 10 minutes per year. Each site must have at least two physical routes between the central master control and each transmitter site. If one path experiences an outage, the network will be required to sense the outage and restore service via the alternate route without human intervention. Alternate route facilities between the central site and each transmitter site will be designated *Main* and *Backup*. Both paths will be subject to monitoring and alarm at all times. The normal mode of operation is Main path on air, and Backup path in standby. If the Main path experiences an outage, the network will be required to switch to *Standby* signal automatically without human intervention. If the Standby path experiences an outage, it will cause an alarm to be initiated and logged. Service restoration will be of the highest priority, and at least equal to any service restoration priority provided to other customers in the class covering public convenience and safety.

Service Provider Technical Qualifications

The service provider shall demonstrate competence through design level expertise with classical and contemporary network technology, systems, facilities, services, and network operations. The successful service provider will own more than 75% of all facilities that the network traverses. For those portions it does not own, it will be required to demonstrate satisfactory contractual business relationships with the owner.

The successful service provider will incorporate network management systems that can be accessed by the buyer for purposes of lodging requests for investigation of alarms and outages. Preference will favor those service providers capable of demonstrating current capabilities to accept requests for investigation of outages or service impairment whereby the user enters information into a terminal or other device with an Internet browser-type interface. It will be an absolute requirement that the successful service provider acknowledge a request within 1 minute of lodging by an automatically generated response, and confirm within certain time frames, depending on priority included in the request.

Responses to this request will be evaluated based on cost and technical trade-offs across performance, robustness, and reliability. Presentation of facts and figures is left to the service provider; however, greater consideration and eventual award will be given to the proposal with meaningful, detailed design characteristics that are related directly to service cost. Highly summarized, the most credible proposal containing best performance, robustness, and reliability constrained by most favorable start-up and long-term cost is most likely to be accepted. The buyer reserves full and final judgment on all matters of the selection and award process and will not under any circumstances pay for cost estimates, proposal preparation, or a similar effort, material, or services except by formal agreement to do so in advance of a formal purchase order.

Due diligence will follow initial selection. Due diligence includes physical examination of a cross-section or example of network elements making up the service provider's core network. Terminals and central offices serving all sites will be examined closely and must include face-to-face meetings with technicians, supervisors, and managers who will be supporting the network. The successful supplier will exhibit confidence in its ability to isolate and resolve issues and concerns of all levels of severity in a competent, logical, and straightforward manner. An examination of power systems, heating, ventilation, air handling, and safety systems and procedures will be conducted.

Network Topology

Network topology will cover sites 1 through N. Service providers initial responses should be based on eight sites as depicted in Figure 7–1. After examination of the initial proposals and selection of two or three finalists, additional sites may be added for revised technical proposals and pricing.

Linking in the final network design will be a mix of round robin and point-to-point links between sites. Fully meshed network design is not technically prohibited; however, cost of such networks is likely not attractive. Furthermore, network management and utilization is unattractive because of operational complexity.

Service providers are requested to study the topology and capacity requirements of the STL network and propose at least two approaches. Each approach should be presented with a spreadsheet

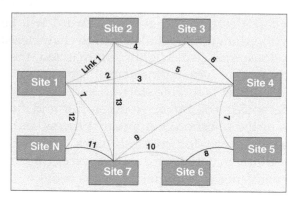

FIGURE 7-1 STL Network Topology

containing link numbers, site designation capacity, and pricing for each link. Specific routing details for each link are required to show active and passive network elements on each alternate route between sites. An analysis showing minimum and maximum time to repair a single failure of each active and passive element in each path or route will be included on a separate spreadsheet in the same file as linking and pricing.

Network Interface and Capacity

For the initial 3-year period, core network interface and capacity needs include diverse STS1 interface (51.8 Mbs) at each transmitter site and diverse OC12/STM4 (622 Mbs) interface at the central site. The service provider will include an estimate of the time, cost, and required notice for making additional capacity available in diverse STS1 increments.

Service interface will be DVB-ASI across a BNC connector on either side of a bulkhead or patch panel of the service provider's choosing. In parallel with each service interface, the service provider will include an Ethernet interface to the network management system to facilitate voice and data communications required to support network management. The service provider may provide access to the public network at their option. If such is included, the service provider must demonstrate adequate and satisfactory firewall capabilities between the network interface point and the outside world.

The service between the central master control and each transmitter site will consist of a minimum of four 20 Mbs channels, net of any overhead, but not including any forward error correction. Forward error correction will be discussed in detail during proposal reviews and a determination made as to its necessity and value.

Premises Architecture and Network Interface

Architecture of the premises equipment includes capabilities to support program content transport, voice, data communications, Internet access, and similar applications such as video conferencing and visual site monitoring. Interface on the network side at SONET/SDH level STS1 or STS4 is preferred. All plesiochronous (PDH) service interfaces will be on the service side of the network interface equipment.

Service Interfaces include SMPTE 259, SMPTE 292, SMPTE 305, DVB-ASI, T1, DS3, and IEEE 10/100/1000BT in accordance with applicable standards. Suppliers are requested to provide a detailed list of applicable standards of official standards bodies such as SMPTE, IETF, ANSI, ITU, or other and relevance and applicability of each reference within the context of each standard's use in their products and services. Figure 7–2 shows a functional block diagram of the general requirements. Potential suppliers are requested to include one or more block diagrams showing the next level of detail along with a matching spreadsheet for each item of equipment proposed.

The spreadsheet must include quantity, description unit price, and total price and discount applicable to each item based on total package price. Separate sheets in the file covering equipment, installation, and maintenance cost during the first and second through the fifth year of operational life are requested.

The service provider may offer network interface device equipment; however, it may not be part of an overall service contract, depending on initial and ongoing maintenance cost. Potential equipment suppliers are encouraged to quote stand-alone equipment with pricing separate from maintenance cost. Preference will be given to potential suppliers who can demonstrate successful, satisfactory business relationships with similar class customers.

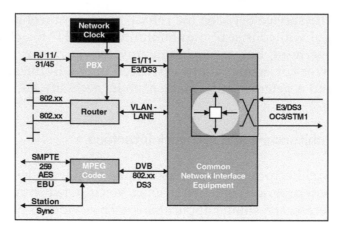

FIGURE 7-2   Premises Equipment Architecture and Network Interface

Design and Installation Services

Services and material of the type required and described in this document are complex and risky. This SOW covers lifting, assembly, and installation of a passive transmission system on and at or near the top of tall tower structures. The major components of the system include a gas stop, transmission line components, hangers, an elbow complex, and an antenna. Successful completion of the work requires specialized technical knowledge and heavy lifting equipment. Preference will be given to service providers deemed in possession of these attributes:

- An established business of providing services such as contemplated
- A strong, accident-free safety record
- The continuous employment of key management and site crew personnel
- A stable financial condition with insurance adequate to cover all risks at each site
- Good relationships with tower designers and manufacturers capable of providing material and design knowledge sufficient to support modifications to existing towers, foundations, and guy cable

- An established relationship with third-party structural engineering firms, or regular employment of professional engineers licensed to do business and practice their profession in the state or states where the work will be performed

SCOPE OF WORK

The scope of work will consist of three phases: planning and design, tower modification, and passive transmission system installation. Specific steps are shown below for each phase.

Phase I: Planning and Design

- Physically examine and survey existing towers.
- Document apparent characteristics, including vertical profile (e.g., is the tower plum?), straightness of legs, foundation, and guy anchor surfaces (e.g., are they level and on the same horizontal plane, and are the guy anchors equidistant around and from the foundation?), an orientation of the tower with respect to magnetic north and the nearest USGS survey or other physical benchmark with direct reference to USGS map. Verify Lat & Lon found in FCC and FAA records for the structure.
- Provide a detailed inventory of all items (e.g., transmission line, antennas, lighting, wiring, and ice shields), and note their location on the tower.
- Document the condition of the tower with respect to any observable surface characteristics (e.g., cracks, missing or peeling paint, bent members, loose bolts or fasteners, clamps, and separated welds that need repair or replacement).
- Examine and provide a description of guy cables and foundation, including current guy tensions and ground conductors.
- Probe the earth around the tower foundation and guy anchors to verify each anchor's physical size and other obtainable attributes.
- Provide a description of the size and material used in tower members.
- Drill or file a small amount of material for lab analysis if appropriate.

- Using any known documentation, including reports provided herewith, build a list of documented references and include in a computer model used for such purposes by the respondent in the ordinary course of business.
- Provide a detailed analysis of the tower showing current load and headroom compared to current EIA or other commonly used standards applicable to the wind load rating at the site(s) tower(s) structure(s).
- Provide a detailed analysis of the tower with gin pole, cables, and other tools while lifting the heaviest load during Phase II and III outlined next. Analysis must show up, down, and any side loading from effects of tag line pull while lifting.
- Provide an analysis that recognizes the weight of the winch platform and any anchors and assume conditions where the winch would lift itself and its anchors or deadweight off the ground. Would the tower be capable of withstanding such load? At what point and under what conditions would the tower collapse?
- Provide a detailed analysis of the tower showing wind load after DTV and pooled communications antenna and transmission line components are added.
- Provide a statement of work including material, labor, and any other considerations required to make modifications to the existing tower(s) prior to installation of the DTV and pooled communications antenna components to ensure the tower remains within the wind load ratings under an incremental load.
- Provide a statement of work including material, labor, and any other considerations required to install and support test of the passive transmission system components supplied by a third party.

Phase II: Tower Modification

- Apply for and obtain any necessary building permits prior to start of work.
- Make repairs and modify the tower in preparation for installation of the passive transmission system components.

Phase III: Install Passive Transmission System Components

- Coordinate and collaborate with passive transmission system supplier.
- Move all necessary lifting and installation tools and equipment on site to unload passive transmission system components.
- Build or otherwise supply wooden support trestles for the antenna.
- Offload the antenna and other equipment from its delivery vehicle(s). Place the antenna on wooden trestles in accordance with the instructions and under supervision of a representative of the manufacturer. If the antenna is a panel antenna or is made up of multiple sub-assemblies, provide labor to assemble the antenna on the ground.
- Install gas stop, horizontal transmission line, bottom elbow, vertical transmission line, all hangers and mounting brackets, tower-top elbow complex, and antenna.
- Support test and evaluation efforts of the passive transmission supplier's representative as outlined below.
- Remove all tools and equipment from the tower structure.
- Re-tension guy cables to design specifications.
- Arrange to have a third party remove all rust and corrosion, and then apply one coat of primer and one or two coats of paint in accordance with FCC and FAA rules and other generally accepted commercial practices.
- Remove all equipment, packing material, trash, and other material from site, repair or level any ground surface disturbance to the original or a similar condition.

Functional Description

The passive transmission system will receive one or more frequency and power-level specific RF input signals and cause the signal to be radiated—"Broadcast." The major components in the system include a gas stop, transmission line, tower top elbow complex, and antenna. The system should be gas tight. The manufacturer will propose a method whereby the system will be pressurized and monitored at a recommended pounds per square inch. If the pressure falls below recommended level, and electrical alarm will be generated at the monitoring unit's output terminals.

Reliance On Services Provider

Reliance on the services provider for their expertise and best business practices as provided to other business, government or non-profit organizations performing public service broadcasting under FCC license. The service provider is expected to use highly skilled and knowledgeable people and standard commercial, off-the-shelf products and material for the services specified in this document.

Throughout the course of any and all action surrounding any RFQ, including procurement, supply, and/or installation of any component or part of the passive transmission system invoking this specification, the right to conduct due diligence on design practices and service processes prior, during, and after completion of the work in the RFQ is reserved. Requirements in this specification are not intended to cause the service provider to do anything outside of its normal business practice. If any requirement in this specification is deemed outside service provider's normal business practice, then the service provider is expected to raise the situation and circumstances immediately upon recognition when and whereupon it can be understood and rationalized through mutual understanding and negotiation.

Coordination and Collaboration

During the course of executing any and all actions surrounding any RFQ invoking this specification, success will depend on the coordinated actions of more than one third party. The buyer is, and at all times will be, solely responsible for coordinating, scheduling, releasing, accepting, and/or approving actions as outlined in any contract resulting from any response to an RFQ where this specification is invoked. The process includes such major items as a notice to proceed with design work, site setup, site clearance, work acceptance, and so on. The buyer encourages communication and open exchange at all times with minimal necessary formality. Any concerns about the disclosure of proprietary or confidential information will be resolved with appropriate Non-Disclosure Agreement(s) between the parties.

Design References and Process

This document is a generic specification, and as such, it is applicable to each RFQ as referenced. Design work begins with the RFQ and carries through and survives into any resulting procurement action or contract. The objective is to do enough design work to allow pricing to be firm and fixed for the period stated in the RFQ. If any item cannot be so priced, then it is to be set aside and identified as subject to change upon completion of final design details, specified date of release, or other specific, detailed explanation of the condition.

Mandatory Design Requirements and Work Flow Processes

Unless specifically agreed to in writing as an exception, this specification requires certain processes be carried out and adhered to during all phases of the work.

The service provider will employ or subcontract a registered professional engineer licensed to practice in the state or jurisdiction where any and all work is carried out.

The Passive Transmission System is a "gas tight" system. Preserving the condition of the surface inside the transmission line, antenna, and other parts of the system is critical to long-term stability and trouble-free operation. The Tower and Erection Services provider must contribute to preservation of positive gas pressure at all times practical during installation. The following are minimum requirements and the responsibility of the service provider on-site supervisor:

- The antenna will be under pressure when it arrives on site. The antenna input terminal will include a gas stop. The gas stop must remain in place until the antenna is safely mounted on the tower and the vertical transmission line and tower top elbow complex are ready for connection to the antenna.

- The transmission line must be assembled from the ground up. The horizontal line will be mounted in a three-point spring hanger suspension arrangement. The vertical line will be mounted with a minimum of two spring hangers attached to mounting brackets on tower members. At the end of each working day or upon work stoppage because of weather, the in-place line will have a cap provided for the purpose of sealing the transmission line. This cap is to be installed anytime work is stopped for more than 2 hours.

- Upon completion of installation of the transmission line and before connection to the tower top elbow complex, a precision terminating load will be attached to the line. The line will be purged with dry air or nitrogen whereby the termination load is not tightened gas tight so as to permit "bleeding" of dry air equal to three times the capacity of the line. Usually this can be done over-night or within a few hours. The Passive Transmission Systems supplier representative will make measurements on the line.

- When the tower-top elbow complex has been installed and connected to the vertical line, the precision load will be moved to the antenna side of the complex, the system pressurized again, and measurements made.

- After the measurements are complete and the antenna is in place, the gas stop will be removed and the final connection of all components made. Depending on the length of time and weather conditions, it may be necessary to purge the system again. Regardless, the system is to be pressurized to one and a half times the recommended operational pressure. The Passive Transmission System supplier representative will make one final set of measurements. If there are no issues, then the service provider will confer with the buyer's representative as a final step before dismantling rigging and other tools.

The following paragraphs are extracted intact from the Passive Transmission System Specification and provided for reference information and guidance to the Tower and Erection Services Provider:

"Any antenna designed and supplied according to this specification will incorporate a unique mounting interface to the supporting structure. The antenna manufacturer is solely responsible for designing the antenna and mounting interface. The antenna manufacturer will exchange design reference drawings and information with the tower and erection services provider. The antenna manufacturer will coordinate a mutually satisfactory mounting interface meeting all applicable EIA, SAF, and/or other applicable standards commonly used in such work by both parties. Design documents necessary to guarantee physical orientation of the vertical and horizontal radiation patterns referenced in the RFQ for each site will be provided to the buyer prior to release to manufacture of the antenna. Any work commenced, including material release, prior to approval by the buyer is at the risk of the antenna manufacturer. Approval of any drawings or other information in respect to this requirement does not relieve the antenna manufacturer/supplier of the responsibility for final orientation of the antenna on the support structure. The antenna manufacturer is encouraged to design unique mounting interface to ensure final placement is in accordance with each site's unique radiation requirements. For example, if a particular antenna is either a tower top mount using a pole in a socket or bolted flange mount, it should have only one way in which to interface with the tower top plate or socket.

"Any antenna designed to comply with this specification will incorporate one or more lifting lugs designed to be an inherent part of the structure through a welding or casting process. Drawings, pictures, illustrations, and design details showing clearly how the antenna is to be attached to lifting cable and tag lines will be provided and subjected to design analysis by third-party erection services providers and structural experts who are qualified to render opinions on safety aspects under all conditions including shipping, handling, installation, operation, or removal. Under no conceivable conditions will this requirement be waived.

"Any antenna made up of panels, feed lines, and mounting bracket sub-assemblies (i.e. not a single mechanical assembly) will follow the same process as outlined in 2 above, except that these conditions will apply at the sub-assembly level, such as a panel and its feed lines and radiators— a power divider/splitter assembly or sub-component as assembled, tested, and shipped from the factory under pressure with gas barriers in place.

"Upon completion of the assembly of an antenna will be fitted with a gas barrier, including pressure indicator and drain cock, and pressurized at its input connection to a level twice the recommended field pressure value. The pressure shall be maintained continously until disassembled in the field for connection to the tower top elbow complex after mounting on the tower. If pressure drops, the leak shall be investigated and fixed prior to further test or installation work.

"Upon completion of assembly and any other tests deemed appropriate and necessary by the manufacturer, the manufacturer shall carry out pattern tests to demonstrate that the finished antenna meets or exceeds the vertical and horizontal patterns invoked in the RFQ. The manufacturer is encouraged to use scale models to reduce cost and test time. If scale models are used, the buyer must review the design process and extent to which they are used and approve or waive any part of the manufacturer's standard full-size pattern test.

"If the transmission line component of the system is greater than 300 feet in overall length or if the line will carry more than one RF signal, the line will be laid out in a single assembly and pressurized at twice the normal recommended level under operation in the field with dry air or nitrogen gas commonly used in the industry. The completed, pressurized assembly characteristic impedance will be optimized to a VSWR of 1.02:1 across an occupied bandwidth consisting of the television channel and any FM signals $+/-$ 10 Mhz above and below the bandwidth occupied by all the specified signals. The signal source and detection equipment will be described and noted with serial numbers, the name of the person making the measurements, their qualifications to do such work, and the dates the work was undertaken and completed with all interruptions in the daily routine noted. The line shall be terminated in a precision load of the same characteristic impedance as the line. The source and detection equipment and load used shall be part of the manufacturer's normal test equipment and its calibration traceable to NBS standards commonly used for such purposes.

"All tower top elbow complex units shall be built, optimized, and tested as a single unit. Upon completion of this process and prior to making ready for shipment, the manufacturer will notify the buyer of this fact and provide evidence that the unit has met or exceeded the agreed-upon specifications. The buyer will examine and approve the unit to be made ready to ship.

"Factory RF Pulse and VSWR measurements must be made and recorded. These will be duplicated in the field. VSWR measurements must be made at intervals of .25 Mhz or less."

DTV Transmitter Specification

This specification covers commerical off-the-shelf television transmitter and related products. The buyer intends to use this document to support design and planning activities necessary to upgrade analog television transmission systems to digital transmission capability. The specifications will be used to define hardware, software, services, and performance in procurement actions and accepting deliverable items as outlined in contracts, service agreements, and purchase orders.

Nothing in this document is intended to commit the buyer to accept technical parameters or levels of performance. The buyer will not pay or commit to pay for any products or services unless and until a valid written contract or purchase order has been executed. Information provided in response to any request is deemed to be representative of products, services, and pricing available from carriers, manufacturers, service providers, software developers, or their authorized representatives.

Unless otherwise noted, items quoted in response to this RFQ are to be products designed and manufactured by the respondent and generally available to a wide range of end-users. The respondent is requested to review the specifications included in this request and propose products that meet or exceed the requirements outlined. It is expected that a respondent will demonstrate competence by providing a list of current users of products being quoted; user configuration and maintenance manuals; design, manufacturing, and test documentation under controlled release; and finally, proving acceptable through the buyers test and acceptance processes. Assertions of ISO Certification are welcome but not considered substitutes.

Manufacturer–Supplier Qualifications

Transmitter useful life expectancy of more than 10 years, perhaps as long as 20 years is anticipated and based on past experience with older generation equipment. Preference will be given manufacturers or suppliers of Transmitter Equipment (as defined in this document) deemed in possession of the following attributes:

- An established Field Engineering and Customer Support organization
- Manufactured and sold the same or similar equipment for at least 10 years
- Will commit to making replacement parts available from a US or North American Free Trade Area depot for at least 10 years after shipment of any equipment ordered in connection with this specification
- Currently offers telephone and Internet parts ordering and knowledge base support 24 × 7
- Established practice in regularly scheduled training seminars for station personnel

Transmitter General Requirements

Highly parallel architecture is preferred. Examples of such architecture are 1 for 1 backup of major components including exciter-modulator, intermediate power amplifier (IPA), high power amplifier (HPA), and all power supplies. Each of these components must be available to substitute for its counterpart upon failure of the unit in service. Ideally, substitution of the backup on failure would be via a monitoring point detecting a failure such as loss of DC operating voltage, output signal, high VSWR, or other similar mechanism. Such activity would also trigger alarms available for remote sensing through transmitter status and control interface. Designs employing high power combining networks to combine the output of high power amplifiers may be an acceptable alternative in specific cases where it's deemed to be cost effective. Figure 7-1 is a functional block diagram showing all major elements of the DTV Transmitter System.

Functional Description

An acceptable transmitter system will generate FCC ATSC Output using ATSC standard input signals and commercial grade power. The transmitter must be capable of being controlled and monitored via status and remote control interface as depicted in Figure 7-3.

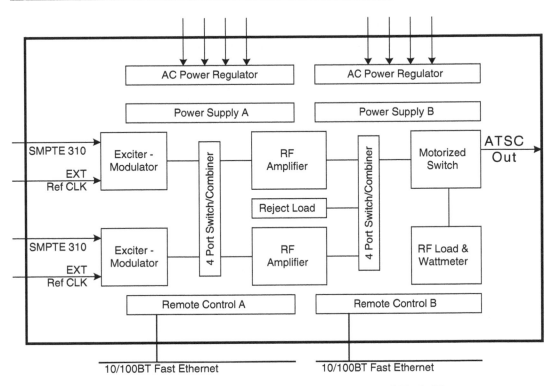

FIGURE 7-3 Digital Television Transmitter Functional Block Diagram

Desired Physical Characteristics

Modular Configuration—Each transmitter shall consist of discrete modules. Each module will be installed in one or more appropriate cabinets, wired, and tested in the manufacturer's plant to minimize assembly at the transmitter site. All transmitter wiring shall be clearly labeled at each termination. The transmitter is expected to include the following or similar modules:

- System Monitoring and Control
- Local and Remote Control via IEEE 802.3 Local Area Network interface
- Uninterruptible Power Supply providing continuous power to the transmitter Local and remote control sub-system. The

transmitter control sub-system will provide complete control and diagnostic information to the LAN interface during absence of power input.

- Exciter and 8VSB modulation
- Intermediate Power Amplifier (if required)
- High power amplifier as required to produce licensed transmitter power output
- RF Switching, filtering, impedance matching, reject, and test loads
- High voltage beam supply
- Calibrated RF power measurement system
- Heat exchanger with redundant coolant pumps and multiple fans for each HPA
- A chain hoist with adequate capacity to change power amplifier tubes
- Interlock system meeting the requirements of IEC-215

Transmitters using liquid cooling will include the following major components:

- Outside heat exchanger, with multiple direct-drive fans
- Redundant coolant pumps with local and remote switching of pumps
- Coolant storage tank with level gauge
- Particle filter
- Pressure gauges
- Temperature and flow sensors
- Test and reject load cooling
- Interlocks for test and reject load, amplifier flow, and reservoir levels

High Power Amplifier Sub-System

The IOT and associated magnet and RF circuit assemblies shall be removable from the front of the transmitter incorporating wheels and quick disconnect RF, electrical, and plumbing fittings. The assembly shall incorporate a positive wheel locking mechanism to prevent accidental movement while in operation.

Electrical, Electronic Characteristics, and Performance

Power Rating—Each transmitter shall be designed for and be capable of operation at the average DTV power specified in the test and acceptance specification.

Metering to permit proper maintenance and roubleshooting, including:

- Collector or Cathode Current
- Collector or Collector-to-Cathode Voltage
- Body Current
- Coolant Temperature
- Output Power
- Reflected Power
- Filament Voltage
- Filament Current
- Focus Magnet Current

High-Power Amplifier Characteristics And Performance

Protection circuitry shall be incorporated to protect the IOT by removing high voltage and drive power from the IOT in the event of an internal tube arc or beam current overload. Any overload condition shall cause the control circuits to remove and automatically reapply high voltage to the amplifier. Three overloads within a short period shall shut down the amplifier and activate the appropriate overload fault indicators. Over-current protection devices shall comply with all specifications of all tube manufacturers approved for the transmitter to avoid violating the manufacturer's warranty. The protective circuitry shall also protect the IOT from other faults that may cause damage to the tube or the tube's circuitry including but not limited to extreme collector temperature, VSWR, and body or grid currents. Supplier shall describe its method of IOT protection and the means by which its tube protection circuitry complies with each approved tube manufacturer's warranty requirements.

IPA/Driver

The IPA/Driver meeting the following requirements:

- Modular design and easily serviceable
- Designed for optimum linearity and fully compliant with ATSC transmission standards
- Rated for at least 120 percent drive power output level
- The IPA and its power supply shall have redundancy incorporated in the design. The IPA shall be provided with an output ferrite circulator for protection against excessive VSWR or an inadvertent disconnection from the IOT amplifier.

Exciter System

Each exciter shall include all necessary switching, control, and status monitoring. The 19.39 Mb/s transport stream input to the exciter will comply with SMPTE 310M or DVB-ASI standard. The exciters shall convert the 19.39 Mb/s transport stream to an 8VSB-modulated carrier. Exciters shall perform frame data randomization, Reed-Solomon encoding, data interleaving, Trellis coding, segment and field sync insertion, pilot insertion and filtering, and be fully compliant with ATSC A/53 Standard and applicable FCC Regulations.

The exciters shall include automatic signal processing for the pre-correction of the signal to compensate for linear and non-linear errors in the transmitter amplifier stages and to provide group delay correction for group delay errors introduced in the output RF system, as well as an RF combiner when present.

RF Output System

The RF system shall be of coaxial and/or waveguide construction, and RF inputs and outputs shall be standard EIA coaxial flanges.

A low loss, constant impedance-type band-pass filter shall be supplied with each transmitter to meet the FCC Mask requirements. The filter shall be supplied with reject and ballast loads.

The RF system shall be supplied with a motorized antenna/load RF switch, liquid cooled test load, and calorimeter power measuring equipment. The switch shall be provided with integral interlocks as a part of the transmitter control system to permit easy connection of the transmitter to the test load.

Two precision calibrated directional couplers shall be provided on the output of the RF system ahead of the antenna/load switch for station use. One indicates forward power, the other reflected power.

High Voltage Supply

The high voltage supply shall be oil filled and self-contained. It shall be designed for operation outdoors over a temperature range of −20 to +45 degrees C (at 100% humidity). Weather resistant panels shall cover all connections. The unit shall be rated for continuous on-air operation at the transmitter's full rated output power plus 20%.

The high voltage supply shall include the capability to adjust the output voltage to the requirements of various tube types and incorporate a step-start device to protect components from large inrush currents.

The high voltage supply shall be capable of operating from the commercial power service at each site. A step-up transformer shall be provided and installed if necessary.

The AC line control cabinet shall control AC power to the beam supply. To protect the IOT amplifier during an overload condition, the AC input to the beam supplies shall be removed in less than 10 milliseconds by high-speed vacuum contactors. The line control cabinet shall include circuit breakers for protection against over-current and short circuits. Mechanical interlocks for both the line-control cabinet and the beam supply shall be incorporated for personnel safety.

Transmitter Control

Overall transmitter system control, monitoring, and diagnostic functions shall be accomplished using an intuitive, high resolution, industrial grade, color Graphical User Interface located in the system control cabinet. GUI screen selection and commands shall be controlled via a touch screen interface. It is desired that basic control functions also be available using hard-wired control buttons located on the control cabinet front panel. These basic functions should include as a minimum; beam voltage on/off, filament on/off, black heat, power raise/lower, and remote/local control.

The following parameters shall be monitored and available for display on the transmitter control/monitoring system:

- Transmitter output power
- Transmitter reflected power
- IPA drive power
- Reject load power
- System interlock status
- All system overloads
- Power supply voltages and currents
- System block diagrams to aid in fault location
- AC line voltages
- Phase loss status
- A summary of active and inactive fault conditions shall be stored and available for view on the transmitter GUI.
- Circuitry shall be provided to lower the transmitter power output in the case of increased VSWR. Decreasing VSWR shall cause the power level to increase until the original output power is restored.
- All logic control memory shall be backed up so as to return the transmitter to its mode of operation immediately preceding an AC power failure of any duration.

RF Combiner

When multiple RF amplifiers are used to meet specific transmitter power output or multiple transmitters operating on different carrier frequencies are fed to a single ended passive transmission system, an

RF Combining until will be employed. In such cases, the RF combiner and all transmitter units shall perform as a *system*. The system will be required to meet all FCC requirements as though it were operating as individual components. The RF combiner shall not degrade the spectral performance of any single transmitter.

The combiner will include an RF sample point at each input and output of the combiner.

The combiner must be capable of continuous operation across the temperature and humidity range as the transmitter with all inputs from all transmitters operating at 110% of their *TPO*.

RF insertion loss shall not exceed 0.4 dB from input terminals to output terminal with all inputs operating at transmitter TPO levels.

Channel isolation shall be greater than 35 dB between any input and any other inputs Group delay shall be within $\pm$ 20 nanoseconds across the channel pass band.

Transmitter System Performance

Each transmitter shall meet or exceed specifications in this section while operating across a range of 25–110% of TPO after installation in accordance with the manufacturers instructions.

TABLE 7-1

Sideband performance	Comply with FCC radiation mask in effect as of date of system acceptance
Harmonic Radiation and spurious emissions	−60 dB RMS, or better
Error Vector Magnitude	4%, or better, measured at the output of an NTSC/ATSC combiner terminated into the test load
In-band Signal to Noise Ratio	27 dB, or better, measured at the output of an NTSC/ATSC combiner terminated into the test load
Pilot Frequency Stability	+/− 200 Hz / month or better. +/−3 Hz or better with external GPS frequency reference
Output Power Stability	+/− 2%, or better.

Line Voltage Regulator

Commercial-off-the-shelf products, such as Staco Energy Products Company, Superior Electric Stabiline WHR Series, or equal AC line voltage regulators, shall be supplied as an integral part of the transmitter design.

User Documentation

All equipment supplied shall include three complete sets of as built drawings, and technical manuals, per transmitter. Each set will consist of printed documents and a CD, DVD ROM, or other computer readable media and include rights to copy, distribute, and use the material for maintenance and education. Content shall include, but is not limited to, installation instructions, operating instructions, tuning instructions, maintenance instructions, theory of operation for all electronic circuits, detailed schematic circuitry diagrams, preventive maintenance, and trouble-shooting procedures. The manuals shall also include parts lists that include the part number, circuit designator, description, and generic number wherever possible. The manuals shall include wiring diagrams with wire numbers and circuit schematics with component designators and values.

The manufacturer shall have a routine policy of providing service bulletins and instruction book updates on all equipment supplied. These service bulletins shall be delivered within five days after release and grant or secure same rights to use as delivered with original documentation.

Acceptance Test and Proof of Performance

A contract for transmitter equipment as defined in this document will also include provisions covering general product proof of conformance to the manufacturers detailed production specifications. Generally, these specifications are the same as included in product brochures or other information. These specifications will be compared to factory product test process and specification limits which may be more, but not less, restrictive.

Initial product shipments covered by this specification will require demonstrated performance in the presence of a representative of the buyer. As experience is gained and the process is shown to produce consistently acceptable results, this requirement may be waived by written notice on a case-by-case basis.

Required factory tests include, but are not limited to, low level tests on the Exciter(s), IPAs, Control Circuitry, and Interlocks. Final test data, including meter readings, dial settings, pads used, and appropriate waveform photos, shall be documented and provided in electronic and paper form for each transmitter tested. The documentation will include all appropriate serial numbers of sub-system components and the transmitter serial number. Name, phone number, and email address of key test personnel, one production supervisor, and one design engineer, knowledgeable of the test process, and results will be included in the test documentation. This documentation will be provided within five days of shipment of each transmitter.

The supplier must provide notification of test date at least 10 days prior to the date the final test process is conducted and test data recorded.

After all transmitter components and subsystems are assembled on site, final acceptance testing shall demonstrate fitness for use and provide data satisfactory for acceptance of the product and formal proof of performance documents bound in a form suitable for FCC License Application. Electronic versions of all documents, forms, and photographs will be required.

The supplier will provide one or more representatives qualified and authorized to represent its interest and participate in the on-site tests and data collection. This test process may be performed by a third party and is viewed as a collaborative effort. Only equipment surviving the on-site test process will be accepted and paid for. Equipment defined as a *transmitter* will be detailed in each RFQ for each site and station. On-site tests will be conducted.

FCC Proof of Performance Measurements required by the FCC for an application for license. Measurements shall be made at the output of the DTV Mask or, if present, the output of the RF combiner. All tests and measurements of transmitter performance shall be conducted with transmitters operating at the power output levels required to meet the effective radiated power specified by the FCC construction permit or license. Power measurements shall be made with the transmitter(s) operating into the dummy load, with results documented as follows:

- ATSC Upper sideband response
- ATSC Lower sideband response
- Envelope delay versus frequency demonstration compliance with Section 73.687(a)(3) and (4) of the FCC Rules and Regulations (47 CFR)
- DTV transmitter frequencies using a frequency counter of adequate accuracy. Measurements shall be made at least three times with a minimum of eight hours between measurements. The frequency reading shall be compared with the reading obtained on the frequency and modulation monitor.
- Spurious components from 0 Hz to 1.8 GHz (if any) apparent in the radiated output of the DTV transmitter. A spectrum analyzer shall be used to make these measurements. Any out-of-band radiation shall not exceed FCC maximum allowable values.

Photographs or other printed facsimile of the waveform shall be taken and shall be included in the proof of performance report.

8

ACQUIRING AND EVALUATING EQUIPMENT AND SERVICES

Management encouraged thinking outside the box. You, or someone, thought outside the box. A proposal was made and approved with enthusiasm by all concerned. Everyone is giddy with excitement. Expectations are high from promises of improved operational capabilities, lower operating cost, and that long list of benefits included in all the sales pitches. Now it's time for someone to sign on the dotted line. Real money is about to be spent. Are you signing or is someone else? Is your job and reputation on the line, or is someone else's? How sure is everyone that all, or at least the most important and valuable benefits or promises and claims of benefits, will actually be realized?

The verbs in the title of this chapter were placed in alphabetical order because that's the grammatically correct way to write. Unfortunately, many Internet and Telecom professionals don't conduct their business in the same fashion. Some get lucky and get by without a major disaster; however, most aren't favored by lady luck and wind up with less of a success than would have been possible had caution not been thrown to the wind.

In the overall scheme of things, communications networks can make or break an organization. In the overall scheme of communications networks, a piece of equipment, software, facility, or service can make, break, or severely handicap an organization. This chapter is about what to do and how to do it to minimize risk and maximize return on investment. Although evaluating equipment and services should be done before acquiring them, it can be done at any time. For purposes of this chapter, it is assumed that evaluation precedes acquisition. Money hasn't been budgeted or allocated, and no commitments have been made.

This chapter is about spending money and getting value in return. It comes in three simple pieces: 1) evaluating potential suppliers; 2) evaluating goods and services against established business requirements; and 3) acquiring equipment, rights to use of facilities, and engaging services.

AGREEMENTS, CONTRACTS, AND TARIFFS

Acquisition of equipment and services is almost always pursuant to or governed by some form of agreement, usually a contract or tariff. Two types of contracts are common: active and passive. An active contract is a form of contract that is not binding on the parties until after it is executed or signed. A passive contract is one whereby the parties are bound upon action of the second party as spelled out in the contract by the first party. An example of a passive contract or agreement is the typical shrink-wrapped software license. These agreements are usually printed on the package containing physical storage such as a floppy disk or compact disc, and state, among other things, that the license is binding on the user when the package is opened.

Tariff is one of the most misunderstood terms in the communications industry. The word tariff dates to Roman times when it applied to tax collection and payment of fees for use of bridges or roads. In more recent times, tariffs were a part of the transportation industry, as well as communications industry. Common carriers included airlines, railroads, trucking, and bus companies. Before deregulation, the Interstate Commerce Commission (ICC) regulated these businesses.

Regulation of the communications industry was and, to a lesser degree, still is the purview of the Federal Communications Commission (FCC) and state Public Utilities Commission (PUC) bodies. Interstate tariff matters are the purview of the FCC and intra-state tariffs are dealt with by the PUC in each state. The PUC also deals with a myriad of other tariffs such as electric and in some cases water, gas, oil, and other material and resources.

More specifically, the term *tariff filing* is the proper term. Entities designated by the FCC and PUC as a common carrier file public documents with the appropriate regulatory body detailing services, equipment, and pricing. The penalty for violation is service disconnection and/or removal of equipment. Tariff filings do not have the force of law. If you or the carrier violates them, then the penalty is service disconnection and/or removal of equipment. Regulatory bodies accept tariff filings, but do not approve, disapprove, validate, or invalidate stated or implied performance, or enforce or police use of the filing.

Attaining capability to order equipment or services delivered by whatever formal agreement, even if it is chosen to order according to a tariff filing, is but a single step in an overall process. Almost any size organization, even if it is a single owner proprietorship, never buys on impulse. This implies some form of practical, prudent due diligence. Conducting due diligence is a key, early step in an overall project. Usually any sizable project will require more than one supplier for equipment, and may well require at least two, and likely three suppliers of facilities and services. Figure 8-1 portrays that process from a point labeled "Bright Idea" through completion of an agreement.

Any real project has a large amount of tasks. And some or many of those tasks can be done in parallel, given the resources. However, there's no getting away from the fact that some tasks can't even be attempted until others are complete. Some can be skipped, or delayed, but the consequences can invite career destruction or worse in the long run. Committing to purchase a piece of equipment without knowing it will function or perform as required, but meets the budget is unwise. Sooner or later, someone in an entity will have to commit to pay

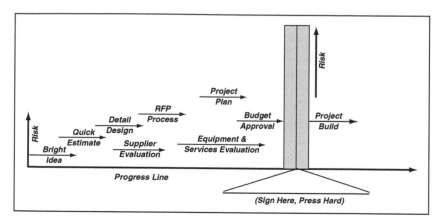

FIGURE 8-1   Project Risk–Reward–Approval Process

before a supplier will agree to ship a piece of equipment, or connect a facility.

The number one overall concern in any project of any size or scale is risk. As a project progresses through the various steps from inception through realization, the number and magnitude of risks associated with or encountered by the business grow and multiply according to scale of the project.

EVALUATING EQUIPMENT AND SERVICES

Evaluating equipment and services is usually within the context of feasibility assessment, or project planning. Risk is minimal and limited to the amount of time and perhaps travel expense invested in this type of activity. As projects grow and commitments are made, risk grows. What we are concerned about is risking capital, and if the opportunity is real, risking the loss of not just the invested capital, but also some or all of the promised return. The worst of all possible nightmares is a scenario where the project causes a problem outside the scope of the project as envisioned, requiring an unplanned, unbudgeted expenditure.

Not only should equipment and services be evaluated, but their source of supply as well. Conducting due diligence on one or more suppliers varies by level of effort depending on several factors such as how long the prospective supplier has been in business, their size, and their capability to produce and deliver the products and services required by the project at hand. A new supplier that's been actively supplying products or services in the marketplace for several years will require less effort than a start-up. A start-up with an innovative, new product may require more effort; certainly it's likely the effort will be different. Well-known or start-up, any situation involving claims of significant new technology should be taken with a grain of salt until all potential material risks have been uncovered, examined, evaluated, and quantified.

Due diligence is most effective and efficient when it is conducted in a way that it becomes a benchmark for acceptance of equipment and services. Conducted properly, there should be no surprises for buyer and seller. If function, form, and performance are not as expected, then there has been a misunderstanding in the past, or something about the hardware, software, or services needs troubleshooting and fixing. Ideally, the due diligence process takes into account the requirements and specifications, and compares capabilities and performance of equipment, facilities, and services of one or more sources of supply. The process should begin with a simple paper-based evaluation. Once everything looks okay on paper, more extensive evaluation can be undertaken.

There are two basic approaches to conducting evaluations. A general evaluation can be undertaken whereby the supplier provides a set of information about the capabilities and performance characteristics of the equipment. The evaluator takes that information and proceeds to determine if the equipment or services are capable of performing as claimed in the information provided. This may or may not determine if the equipment and services meet the business requirements of the buyer. The other approach to evaluating equipment and services takes a specific requirement or set of requirements and then proceeds to determine if the equipment and services meet some or all the details in the requirements established by the buyer.

Once a direction is established, then the process can be sequenced into three or fewer logical steps. Step one can be what's referred to as a paper-based exercise. Simple expert analysis of the information provided by potential suppliers can be analyzed to determine if it meets or exceeds the requirements established by the potential buyer. The next step after paper evaluation requires examination and testing of one or more working samples. Depending on the size and complexity of the project and the potential risks/returns at stake, scalability may be an issue. If scalability is a concern, then the structure of the initial working samples and their evaluation criteria should be staged and structured so as to be extensible to greater scale. Another way to view this issue is to structure the paper evaluation and the initial working sample evaluation so it is representative of the full scope of the network or system as known at a given time.

EVALUATING SOURCES OF SUPPLY

Evaluating sources of supply may be more critical than evaluating what's being supplied. After all, the best equipment or service on the planet is greatly diminished in value if it can't be depended on. Repair, replacement parts, and upgrades are important life cycle extenders. The latest gadget has diminished value if your organization gets sued for patent infringement by the supplier's competitor and the entity the gadget was purchased from isn't around, or doesn't have the resources to defend itself and its customers in such an action. No matter who is right, the outcome is less than satisfactory.

If your organization uses a purchase order process (i.e., limits the power to commit funds to be paid in exchange for goods and services using a purchase order form) then it's likely the individuals authorized to issue POs can conduct supplier due diligence. Initial or preliminary due diligence is general in nature and does not require any technical subject matter expertise beyond a qualified purchasing representative's knowledge of buying what the acquiring organization needs for its business. As project planning proceeds and technical concerns arise, subject matter expertise will be needed to complete the due diligence process.

Evaluating sources of supply is a process whereby accounting, contractual, legal, and technical subject matter expertise investigate and quantify potential risks of doing business with an unknown third party. It involves an examination of the entity's financial health, ability to produce deliverables in the level and amounts required, and its ability and reputation for post-deliverable support of whatever type is required by the acquiring organization.

Evaluating the financial health and stability of a potential supplier can be accomplished by examining a set of audited financial statements covering a period of time in the past. Publicly traded entities that are potential suppliers are required by law to file quarterly and annual financial statements. If the potential supplier is not a publicly traded company, then there may be other sources of financial information such as Dunn & Bradstreet. D&B is a credit rating organization, providing credit history and assessment. As such they have access to information on most any entity that wishes to do business on credit. They also collect information about how the subject pays their bills. For example, do they pay on time, or according to agreed on terms, or are they occasionally or perpetually behind? These same agencies also monitor and report on lawsuits and extraordinary events that may have an impact on the company. Considerations or concerns about the potential suppliers intellectual property rights may be found in simple searches that can turn up patent or copyrights granted.

The overall concern amounts to the ability of the supplier to produce and support the required quantity of equipment, software, or facilities and services for the time required, typically ranging from 2 or 3 years to longer than 10 years. The ability to conduct due diligence isn't rocket science; it's mostly a matter of common business knowledge. You and your peers in accounting that deal with purchasing, contracting, and paying bills can undertake and complete it with a reasonable amount of time and effort. Table 8-1 presents a list of specific steps with a brief description and desired outcome of each.

TABLE 8-1

Technical Due Diligence Process		
Document Requirements	Establish Performance Benchmarks and Capabilities	Provides Objective, Business Oriented Perspective
Request, or Search For Information on Goods and Services	Compare Performance of Goods and Services to Requirements	Produce a List of 2, 3, or More Qualified Potential Suppliers
Create Detailed Technical Specifications	Specify Quantity, and Performance Level Required	Communicate Precise, Detailed Requirements to Potential Suppliers
Create Test and Acceptance Plan	Tells Suppliers How Goods and Services Will be Evaluated and Accepted	Document Formal Acceptance or Rejection Process
Business Due Diligence Process		
Obtain Financial Statements and Information About Size, Customer Base, Market Served	Develop a Profile of the Potential Supplier	Indicates General Capabilities and Limitations
Exchange Terms and Conditions of Sale/Purchase	Facilitates Negotiation of Contract Terms	Creates Formal Relationship Between Parties and Defines How Business is Conducted
Validate Information Provided	Learn from Others' Experience	Assess Potential Behavior and Performance

9

ESTABLISHING AND MAINTAINING RECORDS

How many times have you been asked to help solve a problem or troubleshoot an issue, only to discover minimal or no documentation? Or something less admirable, like the documentation is incorrect or misleading? Fighting a house on fire with a $\frac{3}{4}$-inch garden hose instead of a 3-inch fire hose can limit one's chances of winning. Someone once observed that the difference between craftsmanship and engineering is documentation. Generally, something can't be crafted very well unless it's documented. In complex, large-scale networks and systems, the craftsmanship-documentation cycle is an iterative and ongoing process. There's an initial or draft release used to build up the first working sample. Most times it doesn't work and one or more configuration or connection parameters must be added or changed. Then the document doesn't match reality, so the documentation has to be changed.

Basically, there are two classes of documentation. One class of document contains design reference information and the other is specific to an application or use of equipment, facilities, software, and services. Design reference documents include such things as standards, Product brochures, and specifications, sometimes called *data sheets*.

Another class of documents includes application specific installation, operation, and maintenance instructions supplied with equipment, software, and, occasionally, services. Computer operating systems and applications include some form of *help*. Interactive telephone services based on DTMF tones supporting voice mail, corporate voice network access, and other similar services may be supported with small pocket-sized user guides.

When carriers and Internet service providers build equipment and software into network infrastructure, documentation is critical to operational efficiency and good customer service. It also serves as a list of assets required for accounting records. As you will see later, it's wise to use a single list for accounting, engineering, and operations purposes.

A third type of documentation in the second class is the concern and subject of this chapter. The scope of this type of documentation spans all of the above in terms of its content. Regardless of who does it initially and thereafter maintains it, the potential risk of off-air time may more than justify the up-front expense of its creation. This chapter will define and describe the type, content, and extent of documentation recommended for various size networks and projects.

ESTABLISHING DOCUMENTATION

If you currently use a combination of equipment and network facilities to support content transport, it's very likely that some form of documentation exists. For example, service providers send out bills each month, which are paid by the accounting department. These bills are based on unique circuit identification, usually a series of alpha-numeric characters included in the body of an invoice beside the dollar amount due and payable for use of that circuit or facility. The word *service* is usually included somewhere in the invoice, but what's being paid for in reality is the right to use a specific facility or circuit. If a service is what is actually being billed, there should be a service description or tariff reference identification enabling differentiation between service and right to use of a circuit or facility.

Another example of existing documentation is a circuit identification tag affixed to a wire, cable, patch panel, or network termination point. If these basics do not exist, or cannot be found easily, it would be wise and prudent to request your accounting department and a representative of each service provider establish them.

Existing equipment used in connection with circuits and facilities is owned by the service provider or by the service user. If the service user owns it, then the accounting department should maintain a list of physical assets. If the service provider owns it, it's likely it appears as a separate item on an invoice along with the monthly charge, or a separate invoice altogether.

If an expansion or retraction of an existing network is undertaken, it's opportune and prudent to insist on similar documentation. You can bet the big issue will be "Who does it?" Another safe bet is when you know what hits the fan, it may be difficult or next to impossible to get it done when it's needed most. Like combustion engines need oil and oil filters for long-term, trouble-free operation, content transport networks need documentation for availability, reliability, and robust, error-free operation.

LEVEL OF EFFORT

The level of effort required depends on the size and complexity of the project and the amount of time available or devoted to documentation during the active life of the project. There are two extremes, and obviously an inbetween for all possibilities. Documentation usually begins with a sketch following a bright idea. If carried through to a valuable realization, there are some pretty cut-and-dried steps in between. The process and form of documentation is essentially the same for documenting an existing network as it is for a start-from-scratch case on a blank sheet of paper. Figure 9–1 shows a portrait of what to expect over the life cycle of a project.

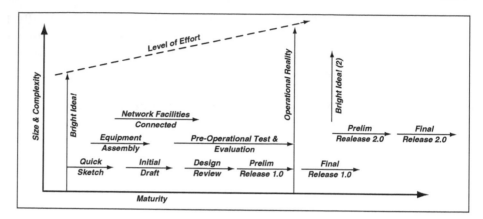

FIGURE 9-1   Level and Content of Documentation Effort

As scope and complexity increase, level of effort increases as the realization of desired operational capability matures.

Approaching documentation this way quantifies it and makes it more acceptable or at least palatable to administrative and management counterparts. It also enhances the ability to manage and control projects, to assess their status and provide milestones for progress reporting and payment. If internal labor is used, and proper accounting for expenditures and transfers between work in progress and completed categories.

KEY DOCUMENTS AND THEIR SOURCES

There are two classes of documentation: design reference information and application or project specific. Design reference documents include such things as official AES, ANSI, IEEE, ITU, SMPTE, or other standards; product brochures, architects and engineer's specifications, construction or building codes; and Federal Communications Commission (FCC) rules. Sources of these documents include the various organizations that exist for the purpose of sponsoring standards development and dissemination: manufacturers, software publishers or their representatives, and various regulatory authorities.

Product- or application-specific documentation includes documents received from or provided by equipment and software suppliers, and documents created by, or for the enterprise, specific to the application.

Application- or project-specific documents include a project description, cost estimate, functional and operational requirements, topology map, list of equipment and services, specifications, and maintenance and operational policies and procedures. Structured properly, these basic documents include budget and schedule information, which can be used to support regular status reviews and reports. It is also wise to develop and maintain some form of a long-range strategic plan, or technology plan. Depending on practice and customs of the particular enterprise, some of these documents may be form based, and fit a particular administrative or technical policy. For example, any project where the capital investment amounts to $1000.00 or more must be subjected to a capital appropriation approval process. Regardless of the amount, the form has boxes for information about the item or items to be acquired, the cost, what they will be used for, justification for the expenditure, and finally, space where one or more executives approve the request. If the amount is large, it may require a resolution for board approval.

DOCUMENTATION DEVELOPMENT PROCESS

Documentation typically starts when a bright idea appears on the scene. The bright idea may fix a problem, expand production capacity, save operating expense, or change the way a company does business. For content transport networks, the project will likely be structured around a combination of equipment, facilities, and services. Carrying the project through to completion will hopefully have a favorable impact on revenue and operating expenses, be profitable when completed, and require significant capital investment.

Starting documentation should be done at the earliest possible time and continue throughout the project with a final release upon completion and commencement of operations. Documentation should reflect reality as closely as possible during the construction and test phases, and especially upon completion. Sloppy or incomplete and uncoordinated documentation leads to mistakes and mismatches, any or all of which can cause re-work, which can be expensive.

Missing details in the final release can impede quick and efficient repair or service restoration.

Documentation should mature as the project matures. Small projects where the project manager functions as the design authority and holds overall responsibility for the project should follow the same process as large projects with dedicated full-time design, scheduling, and financial staff. Components in the documentation include an initial sketch, first draft, preliminary release, one or more revisions, and a final release.

The initial sketch is simply a combination of a brief description of the idea in written form and one or more drawings. It may also include a high-level table or spreadsheet containing preliminary budget estimates of cost, revenue, and profit impact.

A designated project manager should prepare a first draft document as soon as the bright idea gains traction and sufficient interest. This individual should be knowledgeable about the technology and basic business financial aspects of the project. The first draft should be as complete as possible in technical and financial terms. At a minimum, it should contain a technical description, list of equipment, facilities, and services required, a statement of the impact on operational functions, an estimate of the cost, revenue, and profit impact in as much detail as possible, and a schedule or estimated time frames associated with major phases or activities making up the project.

Design review is a critical step in any project. Design reviews are nothing more than a small group of experts from accounting, purchasing, operations, and engineering considering the details of a proposed or in-progress project and making an assessment of the potential risks and unknowns associated with various aspects of the project. A design review can be conducted at any time during the life of the project after the first draft documentation is complete. Depending on the scope and size of the project, it may be valuable to stage multiple reviews as a project progresses. Design reviews may morph into project status reviews over time.

Some practitioners find it valuable to conduct the first design review before release of an RFP, and a follow-on review after responses are received, evaluated, and preliminary vendor selection is completed. Vendor presentations can be made part of a design review where all functional organizations representatives have an opportunity to gain exposure to potential suppliers and assess their ability to perform.

A preliminary release, sometimes called version 1.0, is considered the first example of final documentation. It should include all the documents deemed to be appropriate and necessary for accounting, purchasing, installation, test and acceptance, operations, administration, maintenance, and change procedures. Each document created specifically for the project should contain a reference section where specific third party documents, standards, construction codes, regulatory, and other references are noted. The preliminary release should include updated and extended versions of all documents included in the first draft. Equipment, facilities, and services should be as detailed as possible, and at a minimum include quantity, part number, or other reference identification to be used by accounting when auditing and paying one-time or recurring charges.

One important document in the preliminary release is a formal test plan and acceptance form. This document should be in a state of completion so it can be used to verify physical, functional, and performance characteristics of equipment, software, facilities, and services provided by third parties in fulfillment of purchase orders and accepted proposals. The test plan acceptance form should have matured through the RFP and contract process with each third party. Exceptions or conditions related to acceptance should be documented and any agreement to waive or delay performance as agreed in any contract should be within the context of the test plan.

The documentation should be carefully managed during the construction, test, and acceptance phase. This involves attention to physical details such as cable, patch panel, and rack label details as well as configuration parameters. For example, IP address and phone number details should be checked and double-checked from the time they are planned until final check-off on the completed documentation. Class and type of service designators must be validated.

As the documentation and construction come closer and closer to reality, it may be necessary or appropriate to make additional releases. These should be clearly labeled with a new version number, such as version 1.1, similar designator, dated, and maybe even timed, depending on circumstances and potential requirement to maintain coordinated work to continue. It is also advisable to maintain a change record summarizing the changes included in each release, why changes were necessary, and the specific section and page where the changes were made.

FINAL DOCUMENTATION RELEASE

Final documentation should be revised and updated upon completion of all work and proof of performance of all equipment, software, facilities, and services. Ideally, there should be some form of post mortem akin to the design review that formally examines the operational status, functions, features, and capabilities of the finished project in light of the original design goals outlined in the first drafts and succeeding versions of the overall documentation. Logically this should occur at a point in time where all capital expenditures have been recognized and all increases and reductions in operating expense are known and stabilized.

The final documentation should be archived as well as made available on a website. As such it becomes the reference information for ongoing operations, including paying bills, trouble resolution, service restoration, facilities scheduling, and other day-to-day details required by the business.

CHANGE AND CHURN

Moves, adds, and changes are a way of life in today's business world. Telephones, computers, and other equipment used to support content transport must be in the right place at the right time given the business operation at hand. If the documentation covering the initial project is properly scoped and created, it can support day-to-day operations without change. For example, once a vendor has been qualified and achieves an acceptable performance track record, day-to-day, week-to-week, and other periodic changes can be accommodated with routine procedures.

Changes to the final release should be considered carefully, and only made for a significant change in operations and/or other business requirements. Extending coverage to a new location changes in capacity outside the scope of moves, adds, and changes should be staged and documented with a similar process as used in managing the Preliminary Release during construction, test, and acceptance. Depending on the extent of the change and it's impact on operations, existing documentation can be updated and released as a new release after all the changes are complete and operations are stabilized.

MAJOR CHANGES

Major changes are caused by obsolescence, or outright growth of the business beyond the capability of the preceding design. Occasionally, new regulations bring on unavoidable and sometimes major change. Mergers and acquisitions followed by integration of operations and functions of previously independent entities can cause changes considered major. For whatever reasons, usually consensus of engineering, operations, and finance functions can determine the nature of the change and recommend a path that may require new or greatly revised documentation.

Major changes in the way an entity does business are usually associated with some larger, overriding, and far-reaching event of strategic significance. This event takes on a new name and persona unique and appropriate to the circumstances. As the new circumstances stabilize and the path becomes clear, the implications for existing equipment, software, facilities, and services become clear and planning and design work takes shape to support the new state of the changed entity.

10

ACCEPTING DELIVERED EQUIPMENT AND SERVICES

Acceptance of goods and services is the second and last opportunity to "sign here (press hard)." The first was, of course, when the item or items were ordered. Now its time to see if what was ordered has been delivered. As a milestone in a project, it formally signifies transfer of ownership or right to use the item being accepted. Acceptance can also trigger other internal accounting and operational processes. Acceptance can also initiate warranty coverage, the end of an installation process, and the beginning of maintenance procedures. The acceptance process can be organized, simple, straightforward, and relatively risk-free, or if not approached properly, result in chaos, confusion, and lead to a disaster on some level in the future. Acceptance of equipment and services is but one step in an overall project. The project might be simple or it might be complex. Obviously installation of a telephone or local area network (LAN) access is far simpler than installation of a telephone system or LAN facility. Both examples require acceptance of one or more types of telephone instruments and wall jacks. But a telephone that plugs into a LAN is not the same as a telephone that plugs into a private branch exchange, even though they may both use an RJ45 connector and enable the equivalent and acceptable service.

More than one acceptance process will likely be required depending on the size of the endeavor, project, or organization. Accepting a shipment of 50 boxes containing 10 telephones each without removing enough of them to be convinced that all the boxes have exactly what their outside labels claim can derail telephone service installation and turn up problems in the future if there happens to be the wrong type telephone in one or more of the boxes. Sure, this all sounds like common sense, but the point is to have a process that essentially ensures that what is ordered is received and meets the terms of the agreement covering a commitment to pay for it and take ownership or right to use. Figure 10-1 shows a process whereby a needed facility, product, or service is defined, an order placed (simple case) or contract negotiated (complex case), and the goods and services delivered, accepted, and placed into use.

This chapter will cover acceptance of equipment, facilities, and services. Equipment includes hardware, software, and systems. Facilities include access, switching, and transmission. Services may include items such as professional or consulting services—work-for-hire, or

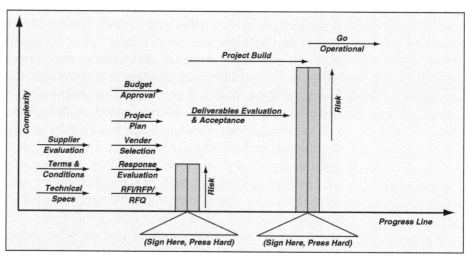

FIGURE 10-1 Ordering, Delivery, and Acceptance Process

other intangibles offered by a telephone company, Internet, or hosting service provider.

Acceptance of goods and services usually takes place during the pre-operational phase of a large project. If not properly planned and organized, it can delay completion and cause operational deadlines to be missed. If it is not carried out thoroughly or skipped, operational effectiveness of the project can easily be degraded or impaired. Ideally, acceptance is a continuation of a logical sequence that began with a well-written technical specification or requirement and proceeded through a chain of several steps, completing the overall process. The complete high-level process includes technical requirements and specifications, business requirements, request for proposal, contract negotiation, order placement, shipping schedule, delivery, acceptance, installation, payment, and, finally, commissioning or placing the item into operational service.

Delivery and acceptance is the second and last opportunity to "sign here, press hard." The overall concern is risk management. First, realize that as a project progresses, risk increases in scope and size. Before the project goes operational, risk is mainly the money invested so far. After a project goes operational, the scope of risk expands to the total business. Another important point to maintain awareness of is the practice of using two types of contracts: active and passive. Active contracts require action on the part of the acceptor in the form of a physical signature or other equivalent approval. Passive contracts include the provision that certain actions—such as the act of opening software, breaking a physical seal on a package, executing an installation script, or executing an application—indicate acceptance. In other words, when the executioner carries out the actions spelled out in the licensing agreement, they become bound by and accept the terms of the license.

IMPLICATIONS OF ACCEPTANCE

Acceptance of an item on a contract or order is generally understood to mean that the item meets the needs and requirements set out by the buyer, clearing the way for payment to the seller. Upon payment, ownership of the item passes from the seller to the buyer. This is the high-level simple view. The more complex reality of managing communications network capital investment and operating expenses is embedded in the facts and details of the agreement of sale. Complicating this picture are two factors. One is the fact that there are three types of items covered in the agreements: equipment, facilities, and services. Second, the value of the items runs the gamut from pennies to millions of dollars per unit, complicated by the possibility of the charges being one-time or recurring on some periodic basis. Managing costs and expenses is the subject of Chapter 11; however, for now it is important to know and not forget that accepting delivery and agreeing to pay not only impacts the cash on hand, but also the organizations financial statements beginning with the next financial reporting period.

It is also important to recognize other related implications of acceptance such as the end of a test and certification period and the start of warranty period. Another example commonly found in software products is the so-called annual support agreement. Most of these are written around delivery and acceptance of the initial software-licensing period. If the agreement is not written that way, it should be.

ACCEPTANCE STRATEGIES

Depending on the level, size, and scope of the organization and project, one or more strategies can be used. A large corporation with a global network may have a laboratory where equipment, systems, and application software are assembled and evaluated. The outcome of such a project and output from the total exercise can substantially reduce the level of examination, test, and report writing during the acceptance process. A phased acceptance in the form of alpha and beta testing can serve new or improved software deployment very well. With phased acceptance, the idea is to prove that the product works as promised and expected. The phased approach can also prove that the deployment plan works as well as how effective the documentation is in terms of supporting users when they have a problem.

Yet another type of system (e.g., a large enterprise wide accounting system) may have to be examined, tested, evaluated, and accepted in one big exercise involving several people dedicated to the task, with cooperation and support from the entire accounting department and all the departmental budget coordinators.

The challenge is to manage risk while at the same time permitting timely installation and start of operations.

ACCEPTANCE OF EQUIPMENT

Digital equipment and systems always include some form of software. After all, digital hardware wouldn't be of much value without the software. Software sometimes goes by other names or euphemisms such as firmware or micro code. Less often, other software such as interpreters and compilers and utilities may be included or supplied with equipment or part of systems. Application or configuration software may be required for the equipment or system to function properly or meet requirements. Applications software might be the product of a third party developer. The possibilities are many, but what is important is what the requirements are or what was included in the order or supply agreement that caused the equipment to be delivered.

The digital television transmitter specification example in Chapter 7 contains detailed specifications and a general approach to test and acceptance. This gives the potential supplier an idea of expected performance and an outline of proving the equipment meets or exceeds the specifications. What is not included is a detailed test plan. A test plan can be written from scratch by the buyer or alternatively provided by the supplier as part of the deliverables, subject to approval by the buyer. Both require significant effort to prepare and validate. Chances are the supplier or the manufacturer already has a detailed test plan for each component or subsystem used in the manufacturing process as well as a system test plan for a complete unit.

Making a manufacturer's system test plan valid for use in accepting the equipment is dependent on emulating the intended use and environment. Essentially, this involves assembling the system as close as possible to the final operational configuration. In most cases this is not a big issue. However if the equipment or system will carry significant volumes of traffic, this may not be possible, except through the use of traffic generators. Again, the main concern is to carry out any and all steps included in the supply agreement. Simply making measurements and validating functions and features outlined in the supplier's published specifications is prudent and helps to uncover and resolve any issues or concerns.

Acceptance of Circuits, Facilities, and Services

On the service provider side, similar concerns and processes exist. The governing approach is either in a tariff or some form of supply agreement. If not explicitly mentioned or called out, then the alternate approach is to use applicable ANSI, IEEE, IETF, ITU, or other applicable recommendations. The process and focus of the effort should be structured around the end-to-end service model block diagram. Initial concern is to make sure connectivity exists by examining and testing the underlying circuit or facility. Validating or accepting a new circuit on an existing facility is less complicated but riskier than accepting a new facility. Regardless, the process is fairly simple and very similar.

If the circuit or facility is intra–LATA (local), only one provider is involved. If it is inter-LATA (long-distance), three service providers are involved: two local exchange carriers and one inter-exchange carrier. If the service is international, the international portion will involve two more physical pieces and a local exchange carrier at the far end.

Circuit acceptance in any case involves a fairly simple process and requires relatively simple test equipment. When the service provider informs you that it is ready for acceptance, simply connect the equipment, loop back the circuit at the far end, and initiate a bit error rate performance test for at least 48 hours—preferably 72 hours. The circuit should perform error free. If there are errors, turn it back over to the provider and ask them to fix it.

Once the circuit or facility is accepted, then remove the bit error rate test equipment and replace it with the operational equipment (multiplexer, server, or whatever) and begin operational tests with the complete system.

11

MANAGING COST OF EQUIPMENT FACILITIES AND SERVICES

Managing cost of equipment, facilities, and services is a critical and ongoing part of managing communications assets and expenses. Communications cost is and has been a growing expense since the first telephone line was installed in a business. Over a relatively short time, telephones spread from the first desk to every desk occupied by an employee to say nothing of conference rooms, halls, and entrance doors. Now everyone has a multi-button, three-color, telephone with three-way calling, multiple lines, caller identification screening, and voice mail jail. Things became a little more complicated when management information systems (MIS) wanted a telephone line and a modem so computers could talk to each other. And then there was the possibility and perceived need for a website, mobile telephones, and wireless Internet access. Communications methods and alternatives abound on all sides from any direction.

Changes in the communications landscape of the mid- to late 1990s have been and continue to be heavily influenced by availability of new technology and emerging standards. Table 11-1 outlines a

TABLE 11-1

Application	Process pre-1966	Process post-2000
Audio/Video Conferencing	Hardly Existed	Internal Equipment and Network or Outside Service Provider
Content Delivery	Analog TV; Paper	Digital TV, Web-Based Content Delivery Networks
Content Distribution	Analog Satellite, DS3 Over PDH	MPEG Over ATM Over SONET/SDH; MPEG Over IP Over SONET/SDH
Corporate Directory	Paper	Part Of Web-Based HR Application
Data Applications	Leased Private Line	Internet Access and Transport
Employment Application	Paper	Web Resume Collection, Automatic Screening
Fax Machine	Dedicated CO Line	Hardly Used Now, Mostly Email And FTP
In/Outbound Voice	Analog DID Trunks	Two Way T1, ISDN Access, VOIP, Cell Phones
Information Systems	Client-Server Applications	Multiple Hosts in Distributed Processing and Network Based Storage Environment
Network Access Technology	Dial Up; Dedicated DS1/E1 Private Line	Migrating to DSL; NxE1/DS1; DS3; Ethernet Over IP Over Cable, SONET/SDH
Network Transport Technology	64 Kbs Voice Channels Over PDH	Migrating to Packet Switching Over Unchannelized SONET/SDH
Order Entry	Computer Terminal	Web-Based Customer Entry
Paycheck Distribution	Paper	Direct Deposit, Electronic Check Stub Delivery
Telephone Equipment	Centrex Or PBX	Migrating To VOIP On LANs

comparison of how things were done before 1996 and after 2000. The change is profound and pervasive.

Broadening the scope of the subject, perhaps it's noteworthy to observe that during the period from 1996 through 2001, the fastest growing and most popular items to be acquired included pagers, mobile telephones, and Internet access. Any enterprise with even one iota of self-confidence and competitive spirit has to have a website. Enterprises adopted the time-honored academic practice of surviving: publish or perish. Money spent on web publishing soared with promise of great savings on product brochures, telephone directories, employee guidebooks, and other works formerly published on paper and transported by physical means. Multiple page documents—agreements, contracts, letters, and similar material clogging fax machines and moving through overnight mail and package delivery services became simple email attachments or file transfers over the Internet at a fraction of the cost of physical transport. The short and simple phrase "change the way we do business" had and still has powerful meaning.

Speed or time to market became the primary rationale for blocking competition and growing market share. Enterprises rushed to field an e-commerce strategy for the sake of being innovative. Many firms underanalyzed, overinvested, and forgot or relegated return on investment (ROI) to the back burner, thinking it was a fruit to be enjoyed somewhere down the information-technology road. Many organizations changed the way they did business, except for the manner in which they contracted and accounted for the cost of communications. The term cost as used here is a technical term used by accountants in cost accounting or accounting for cost of goods (and services) sold.

Fundamentally, all costs in an organization must be accounted for directly or allocated to overhead. Profit or non-profit, all entities must have income to meet expenses or cease to exist. Managing communications cost in many organizations is a line on each department's budget simply labeled *telephone expense*. The typical attitude and approach is simply that it exists, and increases or shrinks each year depending on departmental headcount. If employee population grows, telephone expense grows, if it shrinks, telephone expense shrinks. As for other non-telephone communications expenditures, they are likely to be paid, consolidated, and allocated out to various departments by some process that in effect re-writes and restructures factual circumstances without regard for the matching principle used in accounting.

Technical operations cost centers such as MIS and broadcast operations are something of a special case. Although they will have telephone expense as a line item in their budgets, this covers mainly administrative equipment, facilities, and services. Resources required for content transport and supporting communications related to program and special event production may be budgeted and managed as part of the event or program production budget. Data communications equipment and facilities tied to news gathering, special events, or regular programs may be budgeted and managed as part of an overall MIS department support function or be folded into the cost center accounting of the program or event. Application-specific costs may also be broken out into an individual line item. This means it is budgeted and tracked as a peer line item in a department or cost center. Accounting for expenditures outside the standard departmental cost accounting system requires special treatment in the form of a specific project where each expense item is classified, recorded, and tracked in a stand-alone category in a given period, with final disposition or accounting treatment left for a succeeding period.

If your organization doesn't currently practice communications cost management, don't worry. You can start at anytime with a spreadsheet and a small effort and lever your success into the budgeting, planning, and operating practices of the way your enterprise does business. Potential rewards vary depending on the size of the organization, and, of course, reward doesn't come without investment. Before going too far down the road, recognize that communications cost management is not just paying the telephone bill anymore.

ANATOMY AND PATHOLOGY OF COMMUNICATIONS BILLING AND PAYMENT

Billing for communications equipment, facilities, and services has its roots in telephone service billing. For many years before 1984, AT&T's local operating companies generated the vast majority of telephone bills. Pricing was based on tariffs accepted and filed with the Federal Communications Commission (FCC) and each state Public Utilities Commission (PUC). Basically there were two rate structures, one for business or other organizational entities and another for residential service. Pricing was a combination of a fixed monthly charge to subscribe to local service, plus an incremental charge for calls outside the local exchange area, also called a long distance call. Charges for long distance calls were made each time long distance service was used and calculated on the basis of a minimum charge for the first 3 minutes, plus a per-minute rate for each additional minute the call continued beyond the first 3 minutes. The basic minimum and additional minutes rates were also distance sensitive. The longer the physical distance between the end-points of the connection, the higher the rate. Go back and read this paragraph again.

So far, we have a basis and method for calculating charges for service. Nothing has been mentioned about extra charges for other fees and taxes. We can get by this point fairly quickly by using the simple technique our elected governing brethren use when exercising their awesome responsibility to levy taxes—simply tell the merchant to charge and immediately remit a flat percentage against an item or on the total amount. Maybe we should add on one other little detail. How many taxing entities are there? Hard to say, but we can estimate starting with one federal, 50 state, and thousands of local municipal and county entities who want to charge some small amount for the privilege of paying a telephone bill within the boundaries of the governmental jurisdictions where one lives and works.

Now we are at a point in our anatomy and pathology where we have written a high-level description of the method for arriving at the charges for one or a primary telephone line for 1 month of service. With the magic of simple addition and multiplication on an already complex process, we can contemplate printing and presenting a bill for payment. What does the bill look like? First there's a summary of fixed monthly charges for subscribing to the service, complete with taxes. Then there's the long distance calling summary, complete with details about the time the call started, duration, and charges for each call. For one call, or none, add one more page for long distance service. For the 21st call, add one more page. From the service provider perspective, we now have a streamlined process where we can bill for the majority of our customer base as long as they are single line customers. Oh, and by the way, because we are considering this from the telephone company's perspective, we also have a streamlined provisioning process whereby every single order for changes or new service drives the single line billing process. What happens when someone wants a second telephone line? Simple answer: Install one and send a second bill each month.

By now your hair is hurting, or if it isn't, then get ready, it may begin to hurt shortly. Sure it's painful, but you must understand this picture, else communications cost management will remain elusive and without value. Remember, we are trapped in telephone service billing and payment before 1984, which means one bill for each telephone line, one payment for each bill. Not bad for residential and small business, school, church, library, or other organizational entity to deal with. Telephone service for the masses. What about the larger organization with a corporate headquarters and tens, hundreds, or maybe thousands of field offices? Or a state, county, or large city government with hundreds or thousands of employees in many locations, each with at least one telephone line. You guessed it; phone bills in boxes, not envelopes, measured in pounds, not pages.

The telephone company rose to the challenge, installing service, which works well, and delivering a bill for which payment is expected in due course. Now it's someone's job to validate the bill and pay. And with that, we have arrived at the point where it starts to make sense to consider communications cost management process and practice in large and mid-size organizations—business, non-profit, government secular, or whatever—the consideration is how many telephone lines are required and being paid for. And remember, this is before 1984, and only applies to telephone service, single line access to shared switching facility based services called plain old telephone service (POTS). What's happened since 1984, you might ask. The simple answer is that beginning with the divestiture of AT&T and deregulation of the long distance communications industry, businesses and residential subscribers had to replace single line, single-service billing with triple billing and in many cases initially with triple payment, one for equipment, one for local service, and a third for long distance service. (For more detail on why, how, and exactly when this happened, see Chapter 2.) Stop and think about it—the number of billable items tripled. The amount of envelopes and postage tripled or maybe quadrupled.

Fast forward to 2003 and consider again that communications cost management is not just paying the telephone bill any more. What was a single bill changed into multiple bills because of deregulation and divestiture in 1984. Since then, technological advances and availability of commercial products and services from hundreds or thousands of potential suppliers morphed into hundreds or thousands of bills representing millions of transactions for data traffic, mobile telephones, pagers, two-way radios, remote site security and surveillance services, Internet web access supporting internal and external supplier, and customer transactions, as well as plain old telephone service. Now instead of one trusted supplier whose bill was (appropriately or inappropriately) trusted for so many years, we now have many suppliers requiring some level of validation and due diligence to comply with basic accounting rules, tax statutes, and fiduciary responsibility. Read this paragraph again. Not only are there common sense reasons to manage and pay communications costs wisely, there may be legal implications as well. Check out the Sarbanes-Oxley Bill of 2002 and talk to your attorney and an outside auditor.

CHALLENGES AND POTENTIAL REWARDS OF COMMUNICATIONS COST MANAGEMENT

The accounting question in lay terms: Are the charges valid, can they be verified, and are they treated in a manner consistent with accounting practices across communications expenses and assets?

Beyond basic compliance issues, there are immediate potential hard-dollar savings. There are also other benefits that come from good management practices. Potential savings vary depending on the size of the organization and the effectiveness of current practice. Realizing the savings requires headcount and computer resources appropriate to the size of the organization. To get a feel for potential impact you can compare your organization to large and mid-size business.

Fortune 500 companies average $10 to $12 billion in annual revenue and typically spend $100 to $120 million on communications. Mid-size companies average $18 to $20 million in revenue and spend $20 to $30 million on communications. It takes between one and seven heads to pay the (monthly) telephone bill, which typically ranges between a few hundred and several thousand invoices, where each invoice may contain several hundred or several thousand pages (or pounds). The monthly delivery of the telephone bill becomes a point where the physical work involved in breaking the bill down and entering the actual charges into an accounting system is a daunting effort. Validating the charges is almost never done, and that's one of the areas where cost management can realize significant hard-dollar savings. How much is the issue. General industry consensus based on experience is a range of 10% to 12% of total spending during the first year, tapering to half that level over the next 4 or 5 years. If one took the numbers literally, that would indicate a positive impact on earnings for the Fortune 500 on the order of $10 to $12 million or 10% of earnings, and $2 to $3 million for mid-size companies. How to realize that kind of impact on net income is attention-getting for any sane chief financial officer or chief executive officer.

Very few, if any, large companies receive and pay paper phone bills today. Even with the demise of paper invoices, the number of transactions and invoices cries for some form of automated invoice validation. Pricing complexity for basic services, plus the burdens of taxes, access charges, universal service fees, and other add-on charges imposed by federal, state, and municipal tax authorities all add up to a significant challenge. Adding normal business expansion, contraction, churn, and change doesn't make it easier. Automated invoice verification requires the bill in machine-readable form and an accurate, well maintained reference database containing standard cost and inventory of circuits, equipment, facilities, and services.

Defining the Problem

Recall the earlier description of what is included in telephone service bills—they are highly summarized and comparing each charge in each invoice is very labor intensive. To put the problem and a potential approach to solving it in perspective, it is helpful to diagram it out as shown in Figure 11-1.

On the service provider side, it is easy to see that the billing submitted is made up of many components. The basic service charges include a fixed monthly charge for local service and variable

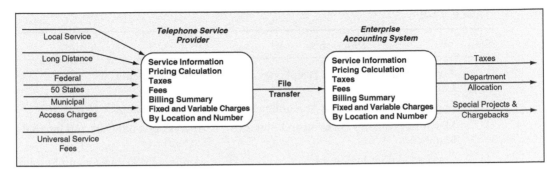

FIGURE 11-1 End-to-End Billing and Cost Allocation Process

charges for long distance service. In addition to basic service charges, there are a myriad of federal, state, and municipal taxes, FCC access charges, universal service fee, and others. (Note: More on the subject in Chapter 12 Budgeting.) FCC access charge funds do not get remitted to the FCC. They go into the RBOC Treasury. This has implications for the future as far as eliminating variable long distance charges.

On the organizational side of the problem, we have two basic needs: keep track of taxes that may have an impact on corporate income or other taxes and fees that may be useful when dealing with community relations or investor information, and allocate total expenditures to all the departments in the accounting system.

WORKING TOWARD A SOLUTION

Obviously, the common-sense approach to this kind of problem is one or more computers and some application-specific software. If you are already performing some form of automated invoice validation, it may be appropriate to review the details of your system to ensure adequate accounting treatment and that all payments made amount to what is due according to applicable contracts, orders, and tariffs (Table 11-2).

This is a worse-case approach, and although few organizations pay paper invoices, it assumes that this is the current practice.

TABLE 11-2

Step	Description	Result
1	Acquire Billing File from Telephone Company	Eliminates Paper
2	Import Desired Data into Accounting System	Eliminates Paper
3	Match Key Fields – Location and Phone Number	Align Phone Bill Data and Internal Reporting System
4	Conduct Circuit Inventory	Identify and Define Circuit Resources
5	Conduct Equipment Inventory	Define Communications Assets
6	Conduct Facility Inventory	Define Capacity and Establish Status
7	Conduct Service Inventory	Define and Establish Approved Services and Options
8	Conduct Contract and Tariff Audit	Establish Approved Pricing as Internal Cost

Automated Invoice Validation System

Automated invoice validation requires a file from the telephone company. Most, if not all, telephone company–provided billing information is assembled in standard formats and made available on various transport media or transported across a network connection. If no validation system exists, or for whatever reason it's desirable to design and build a new one, the process involves getting a file and proceeding to extract the required information from it. The key fields used to match to department allocation are the location and telephone numbers. Information in these fields must be unique. It is highly likely your organization already has unique location names or numbers. In and of itself, the unique name is not sufficient for full allocation. The telephone number making and receiving calls is unique to the telephone company, and they have their own counterpart to the location name, which is a physical port on a switch or access facility into the nearest wire center.

In addition to the key field data, detailed unit count and pricing for all fixed monthly charges and all variable or usage based charges should be extracted and carried over into a file in the organizational accounting system. In addition to the charges for service, taxes and other fees such as access charges and universal service fees should be extracted and maintained in the file. Essentially what has to be done is to take this file with the current charges and perform all the calculations done by the telephone company and re-create the same information, but instead of using the reference pricing they used, use a separate reference database that will be created in steps 4 through 8 of Table 11-2.

Figure 11-2 is a conceptual diagram of the invoice validation system.

It must be implemented in the organization's information technology (IT) infrastructure or built separately and integrated with the other systems.

The functions in this concept are intended to accomplish steps 1, 2, and 3 outlined in Table 11-2. Note the dotted line from vendor

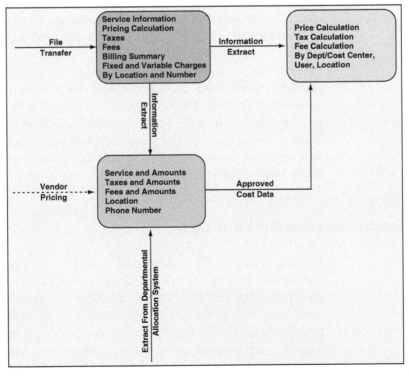

FIGURE 11-2 Telephone Billing File Interface to Enterprise Accounting System

pricing. It will become solid after completion of the standard cost development in steps 4 through 8. Resist temptation to jump too far too fast until you're confident the following functions are being performed:

- Import the billing file
- Extract the desired data
- Compare the data from the reference file with the billing file and highlights exceptions
- Output a list of invoices with amounts due by department, location, and user name satisfactory for feeding the cost allocation process
- Output a file with taxes and desired details required for allocation and further possible tax treatment
- Output a file with any or all fees in a format acceptable for use by the accounting and reporting systems

This will provide a preliminary look at the difference between manual invoice validation and automated validation. So far the only expenditure is the billing tape (should be under $50 for each telephone company, $100 for one local and one long distance) and the value of the time of the programmer(s). When the bill data of the first service provider(s) is successfully imported into the accounting system, incremental service providers, locations, and service should require less time and occur smoothly. If there are issues with the second, be sure to go back and resolve any common issues with the first one. Over time and with additional service providers, adding more billing into the system will become routine.

Establishing Standard Unit Cost

Having successfully established a system to validate invoices using the service provider's billing file, now we are in a position to establish cost reference information. Remember, so far our billing validation used pricing information from the billing tape, not internal cost information. While the two may be the same in theory, in practice and by definition they can't be; otherwise, we have a fox in the henhouse. What we have to do is to establish internal cost information by scouring contracts and tariffs used to acquire equipment, facilities, and services from each service provider or other source in the case of equipment. Most of the required information is likely already in the accounting system somewhere. The challenge, as usual, is to get access to it and then build a routine to deal with it so the result is representative of what's desired. Automated purchasing systems and some of the more recent human resources systems that also contain physical plant information are likely candidates.

Now we come to the point where it may be necessary (in fact it is almost certain) that some type of application-specific software may be required to capture details of communications circuits, equipment, facilities, and services. This is one of the bitter pills often ignored, outsourced, shoved aside, or put into a category of hoping it will go away. If you haven't read Chapter 9 on establishing and maintaining records, it goes into great detail about what should be done, how to get it done, and elaborates in some detail about the critical nature and value of good records. If some sort of system exists, it may be satisfactory. What is needed is an application that can import existing files and is easy to maintain, either from an automatic link to other files and systems, or manually. The output of the system will be used as a reference for the invoice validation function and support other communications cost management functions such as monthly reports, online directory, capacity planning, annual business plan development, and long range strategic planning. Table 11-3 shows the information needed for circuits and facilities.

The information should be collected from documents and then physically verified from tags and labels on terminal blocks and patch panels where available. In the case of virtual circuits and facilities, access to configuration screens on equipment and systems will be necessary to confirm existence and status.

Resources Required

It is possible to outsource invoice validation. There are businesses that are in the business of conducting what they call an *audit* and then using this information to support validation. They also perform other functions such as order administration-new service orders, moves,

TABLE 11-3

Site:		Address:		Main Number:	Contact:
Circuits and Facilities	ID	Type	Carrier	Status	
Inbound Trunks					
Main Number Hunt Group		FAC			
In Line 1		Ckt			
In Line 2		Ckt			
Direct In Dial (DID) Trunks		FAC			
In Line 1		Ckt			
In Line 2		Ckt			
Inbound Toll Free (Inter-LATA)		FAC			
In Line 1		Ckt			
In Line 2		Ckt			
Inbound Toll Free (Intra-LATA)		FAC			
In Line 1					
In Line 2					
Outbound Trunks		FAC			
Outbound Dial (Local)		FAC			
Long Distance Access Facility		FAC			
		FAC			
Point-To-Point Data Service		FAC			
Internet Access		FAC			
Satellite Transponder Services		FAC			
Pagers					
Two-Way Radio					
Satellite Earth Station					
Equipment					
Professional & Consulting					

and changes. It may or may not make sense for your organization to outsource.

Conclusions so far:

- There's value in managing communications cost, and it can be done by third parties for a piece of the savings resulting from outsourcing with the remainder falling directly to pre-tax earnings.

- Effective communications management must be part of the organization's cost accounting system and process; otherwise, significant expenditure of money may lack appropriate fiduciary attention.
- Outsourced, internal, or a mix of all require supervision and informed professional judgment with respect to capturing, recognizing, and managing direct and related communications cost.
- Responsibility and accountability for engaging and supervising the communications cost management resource is an internal executive function irrespective of how it's done or by whom.

COMMUNICATIONS COST MANAGEMENT DEFINED

Managing communications cost and getting satisfactory results may be much more than just paying the telephone bill; however, a stack of paid bills is a good point to start improving cost management. In the typical organization, communications equipment facilities and services are acquired on an incremental, uncoordinated basis, which often results in duplication and decreased utilization. As organizations face the economic reality of changing fortunes and look with favor on time-proven ROI measures and prudent risk assessment, the two-part question: "What are we doing and why are we doing it?" returns to favor. Communications is a dramatically different endeavor than it was 10 years ago. The elements are different and the cost of the desired end result is different, yet many organizations continue managing with the practices and procedures of 10 years ago.

Successfully managing cost in a changing environment begins with, and has its roots in, standard accounting practices and financial planning with appropriate knowledge of communications standards, technology, products, and services. The process involves analysis and understanding of the current situation, followed by consideration of and converting or changing to alternative sources and operations practices.

Managing communications cost is most valuable when conducted as an ongoing internal function. Even if it is, to whatever degree, performed by outside resources, there exist known processes and practices to ensure effective use of organizational assets and resources. The risk is mostly in the level of competence of the people responsible for getting the work done. It is highly likely your organization currently has someone performing certain functions, even if it's only paying the bills.

Resources required to begin improving communications cost include headcount, one or more computers, and some software. For example, several of the companies turning up in the Telecom cost management search on the web offer software specifically designed to validate telephone bills. The software includes built-in pricing references, or it can have specific pricing references installed in it that fit individual enterprise accounting systems and reflect valid charging and rate tables under control of the enterprise, not the telephone company. This software is also capable of receiving the telephone bill from the telephone company or other supplier in electronic form, eliminating the monthly pile of paper forever. There are other capabilities depending on the individual supplier or service provider.

Functions and Work Activities

Executive responsibility for managing communications cost is usually within the MIS or IT organization. Alternatively, it may be within the purview of the executive in charge of physical facilities. The scope of communications cost management fits in the accounting system in similar fashion as physical plant facilities and expenditures for heating, ventilation, air conditioning, water, electric, and power. For example, if the entity has a headquarters in one location and field offices in other locations, communications expenditures will be required to support the people on the payroll at these locations.

Cost Management Functions Deliverables and Results

The next point of interest revolves around specific work activity, deliverables, or output, and the results of the effort or expenditure. For best results, it's suggested that accounts payable and cost accounting functions work in a peer relationship with communications technical expertise. Table 11-4 summarizes the key work items by professional discipline and their attendant value or result.

TABLE 11-4

Highest priority day-to-day short term impact		
Work item	Discipline	Value or result
Invoice validation	CT	Invoice validity
Invoice payment	AP	Liquidate AP obligation
Order administration (moves, adds, changes)	CT, AP	Ensure service when employees move, arrive or depart
Circuit, equipment, facility, and service inventory records maintenance	CT	Vital records accuracy
Unit cost and rate table reference database maintenance	CA	Guarantee accurate and valid price from vendor is accurate cost recorded
Asset tracking and use	CT	Maximize utilization
Medium priority current reporting period impact		
Work item	Discipline	Value or result
Initiate approved capital expenditures and projects	CT	Support business growth
Issue RFPs; evaluate vendors	CT	Enable construction
Supervise and direct contractors	CT	Ensure timely completion of projects
Analysis of spending and resource utilization	CT, CA	Provide basis for change driven decision making
Capacity and resource planning	CT, CA	Maintain service level consistent with organizational growth
Monthly, quarterly, annual reporting	CT, AP	Highly informed and aware management

(continues)

TABLE 11-4 (*continued*)

Lowest priority long-term impact		
Work item	Discipline	Value or result
Participate in departmental and company-wide business plan development	CT, AP, CA	Appropriate and timely adoption of communications technology
Develop and own annual communications business plan	CT, AP, CA	Ensure communications expenditures and resources are managed appropriately and supports overall mission of the organization
Monitor standards development	CT	Advanced knowledge of emerging, stable technology and products
Engage with developers and suppliers of core technology and new products	CT	Provide knowledge of technology and products to support timely adoption into business operation
Develop and own company wide technology adoption plan and capital investment plan that fits overall strategic plan of the organization	CT, AP, CA	Ensure

CT=Communications Technical; AP=Accounts Payable; CA=Cost Accounting

INITIATING COMMUNICATIONS COST MANAGEMENT

Communications cost management (CCM) should be initiated or approved by a senior executive or officer equivalent responsible for ensuring organizational expenditures and accounting are in compliance with accounting and financial rules applicable to the organization. This individual should request or be provided with a briefing on known and suspected level of expenditures directly attributable to telephone service (i.e., data communications, website development, deployment, web publishing, content distribution, delivery, e-commerce, two-way radio, satellite, cable television, and any other service that could be deemed to be or directly support the organization's internal and outside communications).

Communications Cost Management Executive Briefing
Content Preparation

An executive briefing should be prepared for by researching basic facts and parameters of the organization's communications cost accounting and operating practices. The following list of high level actions, analysis, and summary outlines what and how to prepare and present the information:

• Gather all known contracts and billing agreements for voice, data, video, and Internet equipment, facilities, and services. If there are bills being paid according to tariffs on file with the FCC or PUC, obtain a copy of all the tariffs referenced in each billing. If it's convenient and easy to get, include postage meter usage and overnight shipping, especially departments whose work output is not something physical other than paper documents. Every paper document is a candidate for electronic transmission at a fraction of the cost. And don't forget that compact discs and other magnetic or optical media contain files that go quicker through the network at a similar fraction of the cost of paper.

• Gather up at least 3, preferably 6 months worth of paid bills. Each bill should show circuit identification or other deliverable being billed. If the individual items on the invoices are not clear, get an explanation of the item, function, or purpose from the service provider. This should be a written service description that provides greater detail than what is included on the invoice. Usually the invoice references some kind of service description. If the unit price of the deliverable is not included on the invoice or in the service description, request this information from the service provider specific to each invoice.

• Make a list of all billed deliverables paid for and include department, location, and name of the employee using the item. If it is a common item, such as a data line, list all the computer applications using the facility. Accounting might be able to provide a file or printout with this level of detail.

• Find out how the organization acquires or otherwise commits to pay for communications services. The two best places to start are purchasing and accounts payable. Is there a central source for coordinating and consolidating service ordering and terminations? Does each department just order a telephone line or other service or facility when they need it and then pay the bill when it comes in?

• Make a list of all equipment owned or leased and include year acquired, current book value, or unpaid lease obligation, monthly depreciation or lease payment, monthly maintenance cost, and planned replacement date if any. Some or all of this information may be available from accounting asset records or inventory files.

• Summarize all known vendors by service, equipment, and estimated annual unit and dollar volume.

• If cost management is an internal function, summarize headcount, professional specialty, and job responsibility of each individual directly attributable to CCM.

• If a significant level of expenditure is attributable to outsourced or vendor-provided CCM exists, summarize the specific vendors, term of contract, monthly or annual cost, list of work activities and deliverables, and names of key individual contributors responsible for the deliverable, their professional specialty and job title, pay grade, and responsibilities.

The briefing should provide a complete picture of how much money is at stake, what it's being spent for, and the value of the assets attributable to communications functions by department and organizational function. For example, if a set of server hardware and software exists that is used by a single department in engineering, marketing, sales or other area, make sure that department and the function for which it is used is very succinct and clear.

In addition to being clear and succinct about the cost and use of communications assets and expenditures, it is critical that the process and practices being followed be equally clear. This knowledge and information can be used to rationalize next steps and characterize current status. The outcome should be a realistic assessment of the process and practices with respect to compliance with applicable accounting and reporting rules under which the organization operates. Recognition of potential savings is important because real money is at stake. If changes need to be made in operating practices or procedures to improve the competitive or strategic position of the organization, this is the time to get that on the table for consideration as well.

Next Steps

By the time the work in the executive briefing is completed, it is highly likely that additional steps will have been taken, but not included in the briefing. What needs to be done is to take the high level work items listed in Table 11-2, and expand the details of each to gain deeper insight and understanding of how much money is at stake, potential savings, improvement in service levels, better use of human resources and assets, and the limits of flexibility as represented by spare or unused capacity for growth.

Once you have an overall picture of how much money is at stake, and what it's being spent for, you can consider and decide the level and timing of effort appropriate for further work because you are now ready to take constructive, proactive ownership and management of communications cost in the enterprise you're part of. Here's a high-level list of topics and an explanation of what needs to be done, how to go about it, and a view of value or impact on the organization's ability to accomplish its mission.

CONTINUING THROUGH TO OVERALL COMMUNICATIONS COST MANAGEMENT

Machine-based invoice validation is a required first step in successful, long-term continuing cost management. Don't forget the first objective is to reduce expenses and at the same time build a system to maintain good expense control. Overall, the work involves good accounting practices to analyze, classify, and quantify direct and indirect expense and capital investment in areas such as each of the following:

- Cable TV service
- Cable modem service (Internet access)
- Cellular telephone service
- Conferencing services
- Content distribution and delivery
- Fax machines
- Hosting facility use
- Internet access facilities
- Pagers

- Private line data services
- Telephone service
- Website developers

Although it's tempting to make judgments about the value of each while making and completing a list, resist this inclination in favor of focusing on defining the details of how many, how much, how used, and by whom on the first pass. Try to make the list as complete and definitive as possible in terms of equipment, stand-alone application software, facilities, and services. Another important point is to capture the function or functions of each, and do your best to map that function to a business process because sooner or later it will be time to evaluate its impact on the business compared to other expenses or investments.

Get answers to the following questions:

- How are communications circuits, equipment, facilities, and services being used?
- What are the components of each circuit, piece of equipment, facility, or service?
- What are the units of usage and attendant cost?
- What is the volume and total cost of use/ownership?
- Is the cost fixed or variable, and if it's variable what is/are the variable parameters?

Once the answers are placed into a spreadsheet and understood, consider them in light of other questions such as:

- What business are we in, or what's the mission of our enterprise?
- What is each department's role in that mission?
- What is the responsibility of each head on the payroll?
- How does their use, or lack thereof, impact communications cost?
- How can current communications unit cost be reduced?
- How will any considered change impact the ability to deal with suppliers and customers?
- What are the potential risks and rewards from any proposed change?

Satisfactorily addressing the above landscape puts you in a position to consider and undertake several next steps. These include:

- Validate monthly billing (i.e., make sure specific items of equipment, facilities, and services being billed match what is actually in use or in accordance with an appropriate service agreement, contract, order, tariff filing, or whatever caused the billing).
- Make a list of all circuits and facilities terminated in the telephone room. This is a room in each building where telephone company wiring enters the building from the outside. Sometimes there is more than one such room in a single building. More often, one room in a single building serves multiple nearby buildings on the same or directly related owner's property. Prepare the list so it shows unused capacity. For example, in the case of copper cable, the telephone company typically installs what it calls an entrance facility, which is a terminal block on the wall with every single pair of wires in the cable connected to the terminal block. The other end of the cable is terminated on a block in a cabinet, or spliced into a larger cable connecting it to the nearest wire center or controlled environmental vault. In the case of fiber entrance facilities, there will be light wave terminal equipment mounted in a rack or on a wall. This equipment can be used to provision T1, DS3, OC3, or OC12 transmission capacity. Record circuit identification details and get an explanation of the amount of capacity in use and available for future use.
- Compare the circuit list with the items being billed. Many times service is disconnected, but the billing continues. If circuits, facilities, or service is being billed but not delivered (i.e., billing for a fax line, but the line is physically disconnected), you are due a refund. Getting the refund can be challenge without a disconnect order. It is not unusual to find billing errors. Be cautious about use of the term *disconnected*. A telephone line, which may serve or have served a fax machine, may not be disconnected from the service provider's perspective. However, it is an entirely different matter for billing for the line to appear on the bill, and the line not available for use because service was interrupted or discontinued at some point in the past. If a valid disconnect order exists and service was not discontinued, the refund is due.
- Recognize and understand the importance, relationship to, and difference between unit cost and departmental and enterprise wide

expense. How expenses are classified and tracked by a department and then rolled up into the overall picture is important because it's easy to lose sight of duplication and individual item unit cost. For example, it is a common practice to include volume discount from basic pricing in contracts for most all equipment, facilities, and services in use. Don't be surprised if this escapes the attention of the person who writes each service order change or the person who pays the bill each month.

SEVEN STEPS TO PRACTICAL COMMUNICATIONS COST MANAGEMENT

- Invoice validation against standard cost
- Day-to-day order administration—moves, adds, and changes
- Initiating and managing capital projects
- Analysis of spending and resource utilization
- Monthly, quarterly, and annual reporting
- Development of annual budget and business plan
- Development and ownership of technology plan and long-range strategic plan

Invoice Validation Against Standard Cost

The first step in good CCM is validation and verification of all invoices. This requires an accurate, stable cost reference database. This must be established through a process commonly referred to as an audit. Notice that these two steps are independent, yet closely intertwined, and require interaction between accounting, communications technical, procurement, and vendors.

In addition to charges for the service, there are a myriad of service fees, taxes, and other charges imposed on the service by federal, state, and local municipal governments. Further complicating the matter is the fact that contractual agreements covering telephone service are a mix of private one-time agreements, multi-year service contracts, and public tariffs on file with the FCC and state PUCs. Presenting a bill that will be paid on presentation is a balancing act between providing enough information to justify the total amount due and simplifying or summarizing the details so the information will fit onto a reasonable amount of paper or in a reasonable size file, and not seem overly complex.

Telephone service invoices, especially for long distance service, are complex, lengthy, and involve very complex rating, pricing, and tax and other assessments such as access charges and universal service charges. Manual verification of the amounts on the typical number of invoices, and the level of detail is next to impossible. It would be easy to just blame the telephone company for the situation, but that wouldn't be quite accurate because they are required to collect and remit taxes and assessments levied by federal, state, and municipal governments. The telephone company designs and builds billing systems to generate these complex and massive invoices. Given all the complexity, and the fact that the systems evolved over time, mistakes are highly probable, especially with respect to taxes imposed by so many jurisdictions. Successfully coping with this situation is potentially lucrative in terms of credits and reduction of future charges. The most effective way is to fight fire with fire in the sense that it requires well-written application software, significant computing power, accurate rating tables, and constant maintenance of records.

Day-to-Day Order Administration—Moves, Adds, and Changes

Day-to-day moves, adds, and changes may appear simple, but that doesn't mean it is. Across the spectrum, it is in reality like leading an orchestra because it can vary from installing a new telephone for a new employee or contractor to moving entire work groups over a weekend, while at the same time managing multiple contractors in one or more capital construction projects. If the work is not planned and coordinated properly, telephones, LANs, and the entire network service capability can seriously impair business operations. One of the keys to saving money and making sure the invoices are validated and paid promptly is solid technical work planning, using standard processes and dependable sources of supply. The level of effort and resources required to support this step is also dependent upon how many service locations are included, organizational or business growth rate, and the amount of time allowed to properly plan the work.

Initiating and Managing Capital Projects

If communications facilities and services can be considered the life-blood of a business, project implementation has to be the heart and soul of organizational growth. Establishing an operation in a new location after renovation or during new construction starts before the first wall is knocked down or shovel full of dirt is dug. Making sure the wire is in the wall, ceiling, or underground at the right time can cost or save thousands of dollars, and delay the overall project. Making sure the right thing is in the right place at the right time requires good relationships with dependable suppliers, standardized requests for proposals (RFPs), timely evaluation, and due diligence followed by supervision and acceptance of timely deliverables.

Analysis of Spending and Resource Utilization

Knowing where you stand with all facilities and resources in a so-called fast-paced environment requires stopping to analyze where the business is at on a periodic basis. Good management of financial resources requires good budgeting in the first place, but good budgeting doesn't lead to good performance without constant attention and awareness of needs against available resources, spare capacity, and what's in the pipeline for both.

Supplier Evaluation and Selection

Serving a customer is about the only thing more valuable than having strong working relationships with a range of competent, dependable suppliers. Achieving the lowest possible cost is best served with competition. Open, fair, and honest competition happens when suppliers are engaged in a balance between informal give-and-take as well as formal RFPs and organized presentations and proposal responses. Suppliers are where the new technology comes from. Understanding their value chain, timing, and content can provide marketplace intelligence that money can't buy.

Development of Annual Budget and Business Plan

No single department should undertake to strike an annual budget and develop next year's business plan without consideration of current and possible communications products, services, and emerging technology. Communications technical and accounting professionals can provide invaluable insight and knowledge that will allow the line manager and the entire management chain to understand their competitive position and make rational, informed decisions about changing or not changing the way they do business.

In addition, the CCM function should draw up an annual communications plan that covers the organizations initiatives and plans to communicate internally and externally, and show how it will leverage expense and capital across all the functions of the business—sales, marketing, engineering, IT, as well as public and media interests.

Development and Ownership of Technology Plan and Long-Range Strategic Plan

Surviving for the long term is the subject of the last chapter. For now, just realize that the forces acting on your organization include competition, regulation, and technology. Establishing a credible long-term strategic plan requires a sound knowledge of emerging standards and technology. Technology is at the root of the continuing change and generally drives regulatory and business change, especially changes brought on by competition.

A well thought-out technology plan should be developed in between annual business plans and bridge the current business operations cycle by 6 months. For example, if your organization begins its annual business planning cycle in mid-year and obtains management approval in the last month of its fiscal year, then ideal completion of the technology plan should be coincident with the start of the annual business plan. Bridging technology and real-world business planning and execution is key to long-term survival.

TABLE 11-5

Site:	Address:			Main number:			Contact:		
Individual Contributor Services Name	Department	Location	Ext	Local Phone	Long Distance	Pager	Mobile Phone	Internet Access	Calling Card

Site:		Address:		Main number:	Contact:
Circuits and Facilities	ID	Type	Carrier	Status	
Inbound Trunks					
Main Number Hunt Group		FAC			
In Line 1		Ckt			
In Line 2		Ckt			
Direct In Dial (DID) Trunks		FAC			
In Line 1		Ckt			
In Line 2		Ckt			
Inbound Toll Free (Inter-Lata)		FAC			
In Line 1		Ckt			
In Line 2		Ckt			
Inbound Toll Free (Intra-Lata)		FAC			
In Line 1					
In Line 2					
Outbound Trunks		FAC			
Outbound Dial (Local)		FAC			
Long Distance Access Facility		FAC			
		FAC			
Point-to-Point Data Service		FAC			
Internet Access		FAC			
Satellite Transponder Services		FAC			

Equipment & software	Acquired	Cost	Value	Mo dep	Maint	Rep sch
PBX						
Phones						
Voice Processing System						
Auto-Attendant						
Voice Mail						
Interactive Voice Response						
Channel Banks						
Conferencing Equipment						
LAN Equipment						
Content Distribution Equipment						
Content Delivery Equipment						
Mobile Phones						

12

BUDGETING AND FINANCIAL PLANNING

Budgeting and financial planning are not stand-alone activities. Budgeting or making a budget is the act of quantifying cost or pricing for an item or list of items, and in the case of communications, the list is likely to be a combination of goods and services. The items on the list may be a one-time occurrence or recur on some periodic basis. Financial planning is the process of striking several budgets based on alternative scenarios to arrive at a point where the budget fits within an overall set of financial parameters for the organization. The result of the budgeting and planning processes provide a financial benchmark within the overall business plan. The budget part of the business plan includes revenue, expenses, and net income, or if expenses exceed revenue, a loss, which is a big no-no. During the course of the year, the budget will be used to compare actual results of operations in current and previous periods, and to forecast results in future periods.

Organizational budgeting and financial planning for an organization typically rests within the chief financial officer function. There are two activities to be concerned with. Accounting is responsible for entering transactions into the system and periodic reporting of actual results. Successful budgeting and planning requires contribution, collaboration, and cooperation from department managers and subject matter experts as well. The communications cost management process described in Chapter 11 envisions working closely with both groups. Successful communications cost management depends on a high level of consensus among all three groups, and in turn all the department managers are responsible for creating the budget and managing to success and satisfaction of management and stakeholders.

The purpose of this chapter is to provide understanding and insight into creating and planning communications operating expense and capital budgets to fit within an existing accounting and financial operations system. However, significant improvement in managing communications cost (outlined in Chapter 11) is likely to require changes in the way communications expenses are first accounted for and then budgeted in the future.

Recognize up front that changing accounting practices and processes should not to be undertaken lightly because change causes variances when current period performance is compared to previous periods. It can be seen as an inconsistent practice, which is a big flag-raiser with management, auditors, and investors. Getting past the flag requires an explanation. Be prepared up front, and then as the changes take effect and are noticed, be prepared to explain them again. Initially, most of the changes required to realize significant savings involve reclassification of expenses in the telephone expense account. This can be a natural outcome of implementing machine-based invoice validation and establishment of a standard cost system for communications.

Establishing and implementing standard cost and machine-based invoice validation so that each invoice is classified, coded, and recorded prior to being paid provides a basis for analysis, which can in turn support more accurate budgeting and financial planning in the future.

The bottom line is accounting will have to be convinced that the advantages outweigh the disadvantages before changes can be made. There is a two-part message: (1) be convinced and sure of your ground, and (2) be prepared to educate, inform, and convince your colleagues and executives there's "gold in those hills."

ENTERPRISE REPORTING REQUIREMENTS

Reporting requirements vary by organization and are carried out in accordance with laws, rules, regulations, and the organization's charter. Detailed treatment of the subject is outside the scope of this book; however, it is worth a few lines to put communications budgeting and planning in context.

Financial reporting has a public or external face and an internal or proprietary face. Communications financial information is typically 100% internal and proprietary. Hardly ever does any organization make details of its communications matters public. However, as you will see shortly, the information that is released through public channels depends on accurate record-keeping and forecasting to support the accuracy and validity of the higher-level information made public.

Public Reporting and Information Release

Public, or external, information is limited, highly controlled, and subject to rigorous legal and other reviews before being released. Financial reports are released four times a year at the end of each quarter. The basic release includes short factual summaries of revenue, net income, and growth calculations. Unaudited profit and loss, balance sheet, and funds position statements are included with quarterly releases and full audited statements are included in the annual report. Figure 12-1 illustrates how financial information makes its way from information recorded each month through to quarterly and annual reports.

In addition to the quarterly and annual written reports, it has become customary in recent years to communicate financial guidance with respect to expectations for meeting revenue and income growth. Underpinning the public reporting and information structure is an elaborate, detailed accounting and financial planning system.

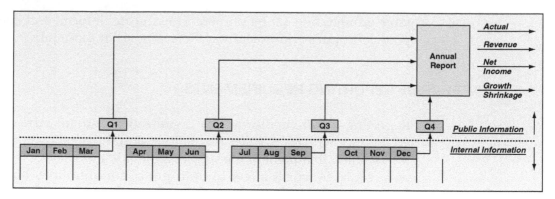

FIGURE 12-1 Information Summary and Flow

A word to the wise: If you have been invited to speak, make a presentation, or appear on a panel at a conference, find and make friends with someone in the group that deals with investors and the media well before the event. Tell them what you've been asked to do and ask for their guidance and support. Chances are they will jump at the chance to support and help you with your speech-writing and presentation charts. The strong advice is to avoid the temptation to not ask for their support and help. When you appear in public and say anything, you are automatically deemed to be a representative of your company, especially so if someone introduces you as "so and so from Company X," or mentions your company at all. If your company is a publicly traded company, you could be deemed to be an insider under securities law. Your media relations or public relations group can help you avoid career destruction, so involve them early and to the extent they are willing and helpful.

INTERNAL REPORTING

The typical enterprise runs on internal monthly reporting of actual results recorded in accounting journals and ledgers. This information is then compared to budget and as the year progresses, a current forecast for future periods. Some organizations use what's referred to as a *rolling forecast*, whereby future quarters and year end coinciding with the annual report is forecast. The rolling forecast is sometimes used to bridge the annual business plan to a long-range strategic plan that picks up from the current year and may cover 3 or more years into the future.

Practices and management style determine how this information is presented and reviewed; however, it's not unusual for an enterprise to detail out an accounting calendar with specific dates each month for a preliminary trial balance, usually within 3 to 5 days of the end of the month, followed by a period of 2 to 3 days for adjustments, and final closing within 10 days after the end of the month. Another common practice is full-fledged senior management reviews shortly after the end-of-month close. This allows release of quarterly results within 30 days of the close of each quarter.

So-called mid-month reviews are supplemented with reviews matching payroll periods. Biweekly or weekly operating results at lower levels in the organization become the time where line management focuses on results measured by orders received (bookings), shipments or installs (billings), orders placed (purchase commitments), and headcount changes (hires and terminations).

Accounting and treasury operations monitor payables, receivables, and cash position on a daily basis, and make decisions about exactly which bills to pay, taking several factors into account each day. Why should a communications manager worry about these issues? Wait until the daily or weekly executive conference call has an echo problem, or security breach, or there aren't enough ports—no, you really don't want to wait, you want to know the calls occur, and you want to know how long they typically last and you want to know if everything works OK, or you want to be in a position to have installed another four or eight ports just before they are needed. As the individual responsible for all communications circuits, equipment, facilities, and services, you need to know about key operating and strategic actions being contemplated and taken. Be aware though that these discussions are highly sensitive and, valuable as they may be, you may not always be invited, or privy to, all decisions until you're instructed to take action or asked to advise on a situation.

ACCOUNTING SYSTEM

The foundation of each organization's accounting system is a *chart of accounts*. This is a name and number system whereby each account has unique identification in the form of a name and number. If a particular department has accounts in more than one physical location, this becomes part of the structure as well. Accounting systems are built around five categories of accounts, assets, liabilities, equity, revenues, and expenses. Table 12-1 explains a little about each.

Within each of these categories are subaccounts used to classify each entry or transaction. If the organization does business or maintains a presence with resources—people, address, buildings, telephone number, etc.—in more than one location, the system will also have location information in the form of a name and number. Understanding this structure is vital to successful management practice in any mid-size or large organization. Communications circuits, equipment facilities, and service are as vital to organizational function as space, heating, air conditioning, ventilation, power, water, and rapidly becoming more vital than postage. Communications is not just paying the telephone bill anymore. Where there was, and still is in most cases, a simple single line on in the expense category there should now be at least three, and perhaps four, for each department or cost center requiring use of communications in its operation.

TABLE 12-1

Assets (1000)	**Items of value owned**. Examples include cash, inventory, and land. Assets are usually tangible, but not always.
Liabilities (2000)	**Debts**. Examples include loans and taxes owed. These debts are usually settled by cash payment, but sometimes they are settled by providing a good or service at a later date
Equity (3000)	**Equity** represents the owners' interest in the company. Equity accounts include common stock and retained earnings or profits reinvested in the business, as opposed to profits distributed to stockholders as dividends
Revenues (4000)	**Revenue** Increases assets when the organizations output is sold or exchanged for money or otherwise valuable goods and/or services. Examples include sales of equipment, merchandise, or voice and data communications, and any interest earned from investments.
Expenses (6000)	**Expenses** decrease assets or increase liabilities when goods and services are acquired and used to fulfill orders and serve customers. There may be literally hundreds of expense categories such as salaries, rent, utilities, advertising, voice and data communications services, taxes, fees, subscriptions to publications, and memberships.

Communications equipment and other assets are accounted for in the asset category. Communications operating expense is accounted for in the expenses category. Accounting for anything to do with communications in the other categories is highly unlikely. If you are, have been, or expect to be given responsibility to manage a department or cost center, you have or will receive monthly summaries of expenses your department incurs. You may or should also know or be aware of investment in or spending for equipment and other assets required for proper operation of the department(s) you're responsible for and that produce the monthly expense summary or summaries. One of the items in the expense summary is depreciation. This is the result of writing off or expensing the value of the asset against revenue over a period of time. It has the effect of reducing taxes and increasing the amount of cash kept from the revenue stream after all other costs are absorbed during each accounting period.

Spending the organization's money begins outside the accounting system when a commitment in the form of a purchase order or proposal acceptance is made. To this day, many organizations still place verbal orders for telephone service. But now even the telephone companies are catching on and it's possible to use their website to place an order for service. Many have had customer order entry systems in place for several years for large accounts. Sooner or later these verbal or computer-based order entry systems cause service delivery and subsequent invoices that are paid and, in most cases, classified by someone in the organization so it fits into the single telephone expenses category.

The expenses category of the accounting system tends to mirror the organizational reporting and management structure. Look at any particular location from headquarters to remote sales offices or stand-alone call centers. What you will see is someone, somewhere in the organization, appointed to be responsible for all expenses incurred during the course of operations at each and every location. Another connection between the system and the organization is in the revenue category and the sales department. Stop and think about it: Is your organization using a website to attract prospective customers and then enabling them to place orders? Where's the sales representative in this picture? The revenue category in the accounting system must be capturing the revenue numbers somehow, somewhere. What does it take to get the expense category to capture the cost of the Internet access facility?

Depending on the size and number of accounts, larger organizations will have additional subaccounts in a hierarchy of sorts that tends to mirror organizational reporting structure and management chain. For example, revenues would look a lot like the sales and marketing organization. Certain parts of the asset accounts would match production or manufacturing organizations where inventory or facilities churn and change on a daily basis as products shipped undergo the billing process. In communications service provider organizations, the equivalent to production is network operations where facilities are used to support the organization's service delivery.

The Account Numbering System

The structure of the monthly reporting system is built around unique identifiers for the account and subaccounts in the accounting system's chart of accounts. At the small end of the scale is the entity that keeps track of its accounting affairs in each of the unique categories: assets, liabilities, equity, revenues, and expenses. At the other end of the spectrum is the large entity with many departments and operational functions located in many locations. At the smaller end of the spectrum are organizations that can easily use two- or three-digit series numbers while the large organization will require three-, four-, or even five-digit account numbers. The same is true for department and location numbers. The key point is to architect the system so that it is expandable and scalable beyond foreseeable growth, while at the same time keeping memory, storage, and processing complexity reasonable.

In addition to the chart of accounts, most modern computer-based accounting systems include, and won't operate properly without, unique departmental identification and location information. For example, the payroll part of the accounting system needs to know which department to allocate payroll expense to, in addition to knowing which employees reside in cities and states with payroll and income taxes.

The combination of account and subaccount identification, department name and number, and location name and number are the basis for communications expenses and capital investment in our mythical organization's budgeting and financial planning examples and practices.

TYPICAL COMMUNICATIONS ACCOUNTING PRACTICE

The concocted examples that follow in this chapter are purposely constructed in an attempt to be representative of mid-size and large company entities with several hundred to tens of thousands of employees operating or doing business in tens, hundreds, or thousands of locations. Many of the characteristics could serve as a template for many media companies in the broadcast industry directly, or others who sell equipment and provide services to them. These examples are intended to be specific to communications expenses and assets across a wide range of organizational entities and types, based on accounting and financial planning practices.

Table 12-2 is an illustration created with fictional (but order of magnitude accurate) data to show an example of departmental expense summary for a mythical entity.

Each department or cost center in the entity will have its own expense summary. If the department conducts operations at multiple physical locations, there may be separate summaries for each location. The example shown is indicative of most communications cost management functions. It is typical of that found in organizations today in that it has a wide range of expense items required by the typical department manager, including headcount and payroll summary, office supplies, training, postage, equipment and software maintenance, etc. The items of concern are telephone expense, equipment, and software maintenance. Most enterprises don't include information about fixed assets in departmental expense summaries. Telephone expense is something of a catchall to the extent that the single line account summary may include depreciation of fixed assets as well as operating expense. This is more of an internal accounting system reporting structure than an accounting treatment issue.

TABLE 12-2

Department: Communications Management   Manager: John Smith   Date: 09/30/03

Department number: 357   Location: 041   Headcount: 8

Expense Category (000 Omitted)	Acct 6000	Sub Acct	Jan	Feb	Mar	Apr	May	Jun	Jul	Aug	Sep	Oct	Nov	Dec	Tot
Salaries		100	47	47	47	47	47	47	47	47	47				424
Benefits		101	16	16	16	16	16	16	16	16	16				140
Incentive Bonus		102	4	4	4	4	4	4	4	4	4				40
Overtime		103	0	0	0	0	0	0	0	0					3
Sub-Total			67	67	67	67	67	67	67	67	67				607
Contract Labor		120													
Professional Services		130													
Office Supplies		210													
Telephone Expense		650	2	2	2	2	2	2	2	2					14
Travel & Entertainment		710													
Training		730													
Tuition Reimbursement		740													
Rent-Space Utilization		780													
Heat, Light & Power		790													
Building Services		795													
Software Maintenance		850													
Equipment Maintenance		880													
Total Department 227			69	69	69	69	69	69	69	69	67	0	0	0	621

Accounting entries are made on a daily basis as operations proceed. At month's end, the books are closed and the process starts again for the following month. These accounting entries are actual amounts of money paid for goods, services, taxes or other items required by the organization to continue in business. For example, a long distance telephone call made in the past and invoiced by the service provider is input into the system and makes its way into the accounts payable for telephone service. When a piece of equipment is received and billed, the information from the invoice is entered into the accounts payable system, and when it's paid for and deemed to be placed into service, the amount makes its way into the equipment section of the assets category, where it's classified by depreciation rules, in turn driving the depreciation account in the (monthly) expense category.

After the monthly closing, various summaries and reports are generated by the accounting system. These reports are used by management to make decisions about buying, selling, hiring, firing, long-range planning, and strategic acquisition or divestiture. Each department manager receives a departmental expense summary covering each department for which he or she is responsible for managing. More specifically, and to the point of financial management, most organizations institute something called *signing authority*. Signing authority is conveyed by management appointment and approval. Senior executives and officers are nominated and approved by the board of directors. Generally, these appointments and authorizations limit the amounts and specific cost centers to a specific list of individuals responsible for approving charges to a specific cost center or small group of cost centers. The individual and their immediate management chain are the only ones who can request and approve expenditure of money charged to the budgets for which they have signing authority. The only possible exceptions are the chief financial officer and chief executive officer, who can generally sign or commit money to be spent for any and all parts of the organization.

Practices will vary by organization; however, it is almost impossible to manage communications without two key summaries. One is the communications cost management department summary of all expenses and assets specific to the communications cost management function, and the other is a summary of just the communications expenses and assets of all departments or cost centers by location in the system. Bear in mind that the accounting system will produce a report for each and every department or cost center, with exactly the same unique subaccount names and number; however, most systems will eliminate subaccount names, numbers, and locations without specific charges to minimize paper and/or storage space. For example, each department with at least one employee will have payroll subaccount names and numbers for salaries, benefits, bonus, overtime, by location, all to whatever level of detail the organization's accounting staff builds and maintains the system. However, any department without any telephones or other communications services expense (Heaven forbid!) would have no communications expense or asset entries. Another example might be training or tuition reimbursement. Unless and until there is an actual expense recorded, there would be no information included in the monthly summary. Once the first entry has been made, it will appear in subsequent monthly reports until the end of the year.

A Better Way to Report on Communications Cost Management

Typical though it may be and grounded in past practice when telephone service was only available from one source like water and power, just paying the telephone bill is far from adequate for effective communications cost management, now and in the future. However, it's a good start and it's not difficult to make incremental improvements over a reasonable time. At a high level, here's what's needed:

• Telephone expense needs to be accounted for in more detailed ways by purchasing and paying for all items of goods and services followed by coding and entry into the accounting system to drive expense reporting categories built around voice, data, paging, Internet access and function specific circuits, equipment, facilities and services such as two-way radio, paging, wireless network access, satellite transponders, and other items that find their way in haphazard fashion into the telephone expense subaccount. Once the basic capability to code, classify, and record the expenses in individual accounts, it's highly likely the reporting side of the system will report in more detail.

- A second, or similar, version of the same form showing total telephone services expense, by department and location. This can provide the communications manager and corporate executives with a clear picture of the level of spending for each category of services across the organization's sphere of operations.

- Addition of an asset category or list of all items of equipment driving depreciation accounts included in the expense summaries, by department or cost center, by location. This information should be equally available and apparent to anyone charged with budgeting, planning, and managing communications cost. Managing the cost of communications is not just paying the telephone bill anymore. It requires capital expenditure, and typically involves consultants and contractors. Moreover, there are strategic implications where call centers and Internet websites sell directly, or support customers.

- Separate administrative equipment and application specific or functional equipment. For example, if a real communications cost management department exists, its administrative equipment and software assets should be budgeted and reported separately from common communications equipment such as voice switches, routers, data switches, network interface devices, local area network (LAN) equipment, etc. This extends to news and other program centric production operations and facilities where communications circuits, equipment, facilities, and services are used to transport content.

- Depreciation expense should be derived from the asset category and given the same classification as basic communications expenses, broken out by voice, data, Internet access, and functional sub-categories.

- Break out of maintenance contracts for software and hardware.

Before launching off into the budgeting and planning waters, let's step back and make a point or two about what has been described and restate the importance of recording communications expenses and assets with sufficient clarity, and detail so they are clear and presentable to the executive bearing responsibility for making sure the organizations money is spent wisely and recorded properly.

Many organizations spend significantly more money than they should for communications equipment and services. Potential savings can range into the tens of millions of dollars or the order of 10% of earnings for the average Fortune 500 company each year. Most executives don't like to hear this; they want to believe they have the best accounting and purchasing available. Others rely on outsourcing solutions or consultants working for contingency fees. Regardless of approach and level of success achieved, it is prudent to examine and review the organizations communications cost management process and practices. Such a review was described in Chapter 11. Properly conducted, it will ensure accurate, valid, proper accounting for and expenditure of the organization's money.

Ideally, there are two versions of the departmental expenses summary. One is reality-based—accounting history—and the other is forward-looking—the approved budget. The rest of this chapter is devoted to describing each and how to build and use both of them to manage communications cost and effectiveness. First we will describe and define the ideal reports, then go into how to use them to get through the budgeting and planning process every organization goes through on an annual basis.

IDEAL MONTHLY REPORTS

The ideal monthly reports will include details about communications operating expenses and fixed assets. The focus of the reports is to declare current reality in the form of actual expenditures to date of the current year, and compare actual results to the business plan or budget. These reports should come from the accounting system directly to each department manager, with copies to each manager's supervisory management chain as determined by organizational policy with respect to dissemination of internal information. Separately, these reports can be used to prepare a forecast that will be fed back to the financial planning group. Examples of suggested report formats will be shown and described. The reports include:

- Departmental expenses and fixed assets: communications management department
- Communications expenses and fixed assets: all departments

Communications Departmental Expense

Table 12-3 is an example of the way a departmental expense summary should appear.

Table 12-4 is an example of a total communications cost summary for all departments or cost centers in an organization. It shows clearly and concisely where the dollars are being spent for manageable cost items.

Table 12-5 is a summary of fixed assets in use by each department or cost center. This report allows communications cost management and individual department managers to see and understand where their investments are and to be reminded once a month about the necessity to plan those investments wisely on an ongoing basis.

It only takes a glance to see the "who, what, where, when" and the level of detail covering the past few months to grasp and see the trends. The principal changes from those shown in Table 12-1 are due to revisions in practices dealing with invoices and the institution of standard cost for circuits, equipment, facilities, and services. Also, note the inclusion of a depreciation line. This is a simple summary; however, it is supplemented by the detail in the new fixed asset report.

This is perhaps an appropriate point to acknowledge that a communications cost management function or department can be staffed internally or outsourced. More likely it will be a combination of both. Whatever the approach, it doesn't reduce the complexity or mitigate the need to account for, budget, and plan. While the operation may vary from organization to organization, there still should be executive responsibility and oversight applied to the increasingly important part of business operations.

COMMUNICATIONS BUDGET AND PLAN STRUCTURE

An organization's annual business plan consists of several components, not just budgets. The purpose of a business plan is to set out goals and objectives—with great focus on financial targets the

TABLE 12-3

Department: Communications Management
Department Number: 357
Manager: John Smith
Location: 041
Headcount: 8
Date: 09/20/03

Expense Category (000 Omitted)	Acct 6000	Sub Acct	Jan	Feb	Mar	Apr	May	Jun	Jul	Aug	Sep	Oct	Nov	Dec	Tot
Salaries		100	47	47	47	47	47	47	47	47	47				424
Benefits		101	16	16	16	16	16	16	16	16	16				140
Incentive Bonus		102	4	4	4	4	4	4	4	4	4				40
Overtime		103	0	0	0	0	0	0	0	0	0				3
Sub-Total			67	67	67	67	67	67	67	67	67				607
Contract Labor		120													
Professional Services		130													
Office Supplies		210													
Communications—Voice		651	1	1	1	1	1	1	1	1	1				9
Communications Data		661	0	0	0	0	0	0	0	0	0				4
Communications—Internet Access		671	0	0	0	0	0	0	0	0	0				3
Communications—Function Specific		681	42	42	42	42	42	42	42	42	42				378
Communications Depreciation		691	6	6	6	6	6	6	6	6	6				54
Travel and Entertainment		710													
Training		730													
Tuition Reimbursement		740													
Rent-Space Utilization		780													
Heat, Light, and Power		790													
Building Services		795													
Software Maintenance		850													
Equipment Maintenance		880													
Total Department 357			117	117	117	117	117	117	117	117	117				1055

TABLE 12-4

Department: Communications Management
Department number: All

Manager: John Smith

Location: 041

Headcount: 8

Date: 09/20/03

Expense Category (000 Omitted)	Acct	Sub Acct	Jan	Feb	Mar	Apr	May	Jun	Jul	Aug	Sep	Oct	Nov	Dec	Tot
	6000														
Contract Labor		120	2	1.5	1.8	15.3	10.5	1.6	1.9	0.8	0.5				36
Professional Services		130	22	9	0	0	0	0	8	4					43
Communications—Voice		651	126	123	131	128	129	127	133	138	139				1174
Communication—Data		661	105	104	104	106	110	111	111	111	111				973
Communications—Internet Access		671	27	27	27	27	27	31	31	31	31				259
Communications—Function Specific		681	49	49	49	49	49	49	53	57	57				461
Communications Depreciation		691	6	6	6	6	6	6	6	6	6				54
Software Maintenance		850	12.5	12.5	12.5	12.5	12.5	12.5	12.5	14.5	14.5				117
Equipment Maintenance		880	18.5	18.5	18.5	18.5	18.5	18.5	18.5	19.3	19.3				168
Depreciation (Detail in Asset Category)		910	6.02	6.02	6.02	6.02	6.02	6.02	6.02	6.02	6.02				54
Total All Departments and Locations			374	357	355	368	369	363	381	388	384				3339

TABLE 12-5

Department: Communications Management   Manager: John Smith   Date: 09/20/03
Department number: 227   Location: 041   Headcount: 8
Fixed asset summary

Equipment	Acct	Sub Acct	Acquired Date	Purch Price		Useful Life-Yrs	Age Years		Monthly Dep	Book Value	Rem Life
Voice Switching Equipment	1000	1640	9/30/2000	60.0	5	3	3	2	1	23.8	2
Voice Processing System		651	9/30/2000	25.0	5	3	3	2	0.42	9.9	2
Voice Wiring		652	9/30/2000	5.0	5	3	3	2	0.08	2.0	2
Telephones		653	9/30/2000	30.0	5	3	3	2	0.50	11.9	2
Sub-Total		654		120.0					2.00	47.7	
Network Interface Equipment		650	9/30/2000	100.0	5	3	3	2	1.67	39.7	2
Uninterruptible Power Systems		650	9/30/2000	5.0	5	3	3	2	0.08	2.0	2
Data—LAN Equipment		650	9/30/2000	28.0	5	3	3	2	0.47	11.1	2
Streaming Server		661	9/30/2000	10.0	5	3	3	2	0.17	4.0	2
Sub-Total				143.0					2.38	56.8	
Program Content Transport			9/30/2000	180.0	10	3	3	7	1.50	125.8	7
Administrative Equipment			9/30/2000	8.0	5	3	3	2	0.13	3.2	2
Total		910		451.0					6.02	233.4	

organization hopes to achieve in the coming year. Typical practice is to begin working on next year's business plan and budget in the mid- to late third quarter of the fiscal year. For organizations operating concurrent fiscal and calendar years, school, and budgeting start around the same time. Usually the business plan and budget goes to the board or owner in December. Well-laid plans get approved; lousy plans become the root of contention and revision during the holiday season. Most organizations don't enter a new fiscal year without a board-approved operating plan for the next year.

Communications budgeting and planning involves only the expense and asset accounts in the system. This is the time and place to get new inputs into budgeting and operating practices, as well as organize new parameters in operating and capital expenditures. Regardless of reporting structure, it is strongly recommended that communications cost management be accounted for as a departmental peer to other operating functions such as sales, marketing, accounting, management information systems/information technology (MIS/IT), etc. Alternatively, it can be part of MIS/IT if all subaccounts are properly structured and their entries classified appropriately. But care must be taken to keep the two separated and well defined, because of the potential for, and sometimes outright, duplication of resources or empire building. Missing an opportunity to acquire a new operational capability is also possible as well. This is the syndrome known as the left hand not knowing what the right hand is doing.

Good budgeting and planning practice uses previous years' actual results as a foundation on which to build the next year's plan. Overall, the process involves studying and understanding previous years' history, followed by development and analysis of alternative scenarios. Reports of actual expenditures by category, department, and location covering the past year are the critical starting point. These reports should come from accounting on a regular monthly basis.

Excellence in budgeting and planning practice dovetails with and leverages successful long-range business and strategic planning. (That's the subject of Chapter 13. For now, focus on current quarterly and annual operations planning.) Gaining a detailed understanding of the content of goods and services making up the numbers may require examination and study of the invoices and contracts that caused the numbers. This is the area where communications subject matter expertise can greatly enhance clarity and meaning with respect to cause and effect of capital and operating expenditures, really purchasing decisions, on individual departments as well as the overall operations of the enterprise. It is important to determine the value of each and every spending transaction. What is the result of providing every single employee with a telephone? What would happen if they didn't have a telephone, or if they had to share a telephone with another person? Managers in all departments with responsibility need to evaluate and consider the work content of each and every employee. What is their input and output; how much is dependent on 24/7 availability of a telephone for incoming and outgoing calls? The same questions should be asked about their computers, LAN usage, pagers, mobile phones, and other gadgets. Not that they aren't valuable, but it's a simple matter of understanding how valuable. And if the value is real, is it being applied to, or used by, all appropriate headcount?

Capital spending should be scrutinized as well. Look at the previous 2 or 3 years of capital spending. What was the cost of each component in the spending package? What was the expected result? Capital spending should either result in savings or profitable revenue growth, preferably both if possible. What was the promised return compared to reality today? Get numbers, because you will (or should be) asked. If you're not asked, then you should update your resume and watch for an opportunity to move to a job where management asks before you're forced to because the management you work for might not get supervised by the bank or board before the business isn't a business any longer.

The budget is only one part of an overall communications plan. Depending on the way the enterprise is organized and conducts its accounting practices, communications budgets, and operating activities may be centralized or decentralized. They may be wholly an internal function or completely outsourced. In reality, it's highly likely somewhere in between the two extremes and a mix of both. Other key parts of the plan include people and vendor or supplier resources. Communications expense and capital expenditures are significant dollar amounts. We know from experience that opportunities for significant one-time and ongoing savings exist. It is not unusual to realize 8% to 10%, or even 15%, favorable impact on pre-tax profit. Budgeting and financial planning time is the time to create a plan to realize those savings.

In addition to impacting the cost of communications directly, budgeting and financial planning provide opportunities to impact the organizations overall growth rate and the competitive and strategic position in the marketplace. Creation of a website with adequate, but not oversized communications network access and just the right amount of advertising and promotion can be just the ticket for a newly created product or service offering. Expansion of customer support with a new call center located in an area where labor cost is lower is a no-brainer. But structuring the design of the communications network and system required to support scalable growth over a 2-, 3-, or 5-year period requires knowledge of communications technology and commercial products and services to design, build, and operate in a way that enables and does not constrain growth. Competent communications budgeting and planning supports the department with direct responsibility to determine and plan the website. The responsible department describes what they want in the way of capabilities and results; communications management designs the facility and prepares a detailed operating and capital project plan, including budgets for both.

BUDGETING AND PLANNING FORMS AND PRACTICES

If you're an experienced budgeting and planning practitioner, it's likely you use a system that has evolved from a combination of learning by doing and classroom training. However, based on what we learned in Chapters 11 and so far in Chapter 12, it seems logical that we should be able to construct a system by simply incrementing it with forward-looking details and minor caption changes.

If we hang on to the intuition to take history and project it into the future, it should be possible to create a foundation from which to make changes and iterate results until we arrive at a point where we have the following:

- Departmental expense budget for communications cost management (CCM)
- Fixed asset account, capital budget, and revised depreciation schedule

Now that we have determined where we want to go, the issue and focus is on how to get there. Experienced budget crunchers have come to learn that once a year, somewhere out of the blue comes a set of instructions and assumptions to prepare the budget. And thereupon begins an annual ritual akin to something between a Mexican hat dance and an Indian snake charmer festival. Typically, the only thing constant, clear, and repetitive is "don't spend any more than we have to," or "cut capital spending by 25%."

Well-run, successful organizations breeze through the cycle and complete the annual business plan and budgeting with aplomb akin to a well-run ballet troupe. Why and how do they do this? Simple:

- They have a common-sense, well-adapted internal reporting and forecasting process
- They get management direction in the form of two or three alternative growth scenarios for the next year
- They use clear and simple assumptions regarding availability and use of headcount and capital
- They operate on a no-fear, no-cut, schedule with dates and deadlines for actions by key players

Before undertaking budgeting and planning work, it may help to explain a little more about what's been going on during the current year so far. Earlier there was a mention of timing of the budgeting and financial planning process. Figure 12-2 shows key elements and timing with respect to the annual business plan.

Basically, there are four activities taking place during the course of an operational year, sometimes called fiscal year. The well-managed organization begins the year with an approved budget and business plan. Throughout the year, operating results are recorded and reported on internally and externally. Even though the plan is fixed and doesn't change, operations and results will vary because of several reasons. Moreover, the organization that doesn't change its

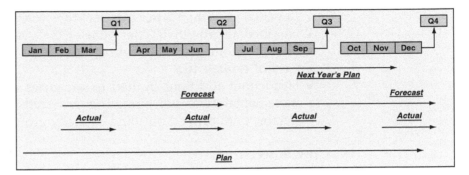

FIGURE 12-2 Financial Reporting Planning Activities and Time Line

way of operating during the course of the year isn't long-lived. At the root of change is the forecast activity. Properly carried out, forecasting is a powerful tool for driving annual operating and strategic business plans.

Assumptions and Growth Scenarios

First of all, let's assume for purposes of the exercise, that we are on the receiving end of the assumptions and scenarios. The effort will involve:

- Responding to the request for a plan and budget
- Preparing a budget for the communications cost management function
- Supporting all other departments with communications expertise in the preparation of their operating and capital budgets.

Here are the assumptions provided each department:

- General economic growth remains sluggish to a point or two on the upside
- Industry segment growth: 3% overall
- Inflation between 3% and 4%
- Business growth in accordance with long-range strategic plan
- Delay replacement equipment capital from first to second quarter
- Delay expansion capital from second to third quarter

- Revenue growth: 5% per quarter, 15% year-to-year
- Net income growth: 6% per quarter, 20% year-to-year
- Short term interest rates: remain under 5%
- Cost of capital: 10%
- Headcount additions limited to vacancies in existing positions; incremental revenue; operating cost reduction projects (contractor, until proved out) and capacity growth.

Growth scenarios:

- Expand regional programming from one currently to two or three areas.

One of the more often ill-practiced parts of business planning is making a plan-for-a-plan, complete with dates and deadlines. This is senior management responsibility. But if it isn't practiced well, and you're the communications manager that has to live with the situation, you can make your own plan and deadlines and benefit from such action. Table 12-6 shows an example of how to lay out an overall schedule and plan.

Budgeting and Business Plan Schedule

In addition to the information above, each department manager has been provided with current financial summaries and first draft expense and capital budget forms or spreadsheets to use in developing the first draft submission.

TABLE 12-6

Step	Start Date	Deadline
Management issues guidelines and assumptions; requests draft plans	August 1	September 1
Management review cycle	September 1	October 1
Revision and negotiation	October 1	November 1
Prepare final plan	November 1	December 1
Final approval cycle	December 1	December 15
Distribute plan	December 15	December 31

SUCCESSFUL PLANNING

Successful planning is a mixture of paradoxes at best. First, it's not possible to judge success until it's history, or as Yogi Berra would say, "It ain't over 'til it's over." And once it's over, we can look back and judge failure, success, or in between. Second, regardless of the circumstances, it's not possible to go back and change what happened or didn't happen. History might be re-written, or written differently, but if it is, then it isn't accurate. Lastly, it's not possible to predict the future, except in terms of probabilities, such as the typical weather forecaster might say "a 20% chance of rain," or "temperatures tomorrow are going to drop into the 80s." In the instant case, it's not possible to predict or know if our planning has been successful until it's too late to do something about it.

Successful budgeting and financial planning is one of the keys to long-term success. If your department or organization doesn't have a plan, then make a plan to make a plan and follow through. You will feel better and improve your chances of success in doing what you want to do. If your organization has an annual business plan, then live by it and as time passes, make adjustments in your day-to-day or month-to-month work to stay on track, adjust or make up for misses, and watch for opportunities to capitalize on. At the end of the year, you can look back and say what you did well, and what needs improvement. Hopefully, along the way, one of the accomplishments was a business plan for the next year.

Once successful budgeting and financial planning has been instituted and seems to be working well for a period of 2 or more years, consideration should be given to longer range planning cycles, sometimes called strategic planning. Large, successful organizations aren't born, they are made—not overnight—but over many nights and sometimes years. It's also true that they didn't get to their present state without mistakes. Sometimes those mistakes were planning mistakes; other times, they were operating mistakes.

One other important ingredient, the subject of Chapter 13, is a long-range strategic plan that dovetails with the annual business plan and well-oiled periodic reporting practices. This last part is no small matter; very few organizations practice it well, most are mediocre at it or simply don't practice it at all.

13

TECHNOLOGY FORECASTING AND STRATEGIC PLANNING

During the weeks before the Cuban missile crisis in 1962, there were many meetings of the National Security Council. The story made the rounds of the staff and later appeared in print about one of the meetings that took place just before the situation was made public. General Curtis LeMay, a high-ranking Air Force officer previously in attendance, was missing. One of the President's aides was concerned and asked, "Where's General LeMay?" The President ignored the aide and continued with his opening remarks, but after a minute or so, believing the General's presence to be critically important, the aide persisted with an interruption: "Excuse me, Mr. President, General LeMay isn't here, shouldn't we wait for him?" At this point Kennedy is said to have stopped, removed his glasses, looked straight at the aide, and said, "No. We don't want him here. We're here to decide whether or not we want to bomb Cuba. If we decide we want to bomb Cuba, we'll put Curtis in the lead airplane, but we don't want him helping us make that decision." This story is sometimes used to illustrate the difference between strategy and tactics. The strategy involved deciding what to do. However it's likely the act of excluding the General was a tactic aimed a keeping minds open and forcing a war council to consider one or more alternatives and forge a strategy or two to get the missiles removed from the Caribbean island.

In the late 1960s, the Radio Corporation of America changed its corporate name to simply RCA and went on an acquisition spree buying Cornet Carpets, Banquet Foods, and Hertz Enterprises. Many saw this jokingly as changing the meaning of the company name to "rugs, chickens, and automobiles." Later, after struggling for several years to compete with IBM in mainframe computer design and manufacturing, the company decided to exit the business. After the decision was finally made and formally announced, an observer noted, "Strategic planning at RCA consisted solely of deciding what to do after lunch."

CONTEXT OF TECHNOLOGY FORECASTING AND STRATEGIC PLANNING

Why should something like technology forecasting and long-range strategic planning be part of communications (cost) management? Because communications is the lifeblood of any organization, and if the blood gets clogged, slows down, or stops, the organization will cease to exist. Technological forecasting and long-range planning assures that communications capabilities and facilities continue to change and evolve to meet the needs of a growing organization operating in an environment subject to change brought on by competitive pressure and regulatory and technological change.

Technology forecasting and strategic planning should be considered in the broader context of the organization's mission and objectives. It should deal with all fundamental requirements of the organization, such as revenue producing products and/or services structure, human and capital resources, physical facilities, and utilities. This chapter introduces the concept of long-range strategic business planning, explains how to do communications technology and long-range planning, and its use in the overall plan.

Long-Range Strategic Planning

Does your organization have a long-range business plan, also called long-range strategic plan, or simply just a strategic plan? Have you seen it? If the answer to both questions is yes, consider your good fortune in the sense that most companies don't have a long-range plan, let alone a technology plan. Much like the story about RCA, they worry about what to after lunch, eventually getting bought up, or taken over by another organization. Others take forever to get an annual budget prepared and agreed to by their banks and other lenders whose primary interest is that they receive each month's interest payment on the money the company owes, not on the company's product and service revenue, gross margin, and net income.

If your organization does not have a long-range plan, regardless of the nature and character of the annual business plan or operating budget, you are in a good position to learn while doing something valuable. If your colleagues and management do not appreciate the effort, you will be in a position to benefit your next employer when your current organization experiences its premature demise. The best place to start is with your organization's annual business plan and budget, roughly where we left off in Chapter 12.

STRATEGIC PLAN: A DEFINITION

Webster's *New World Dictionary* has several definitions of *strategic*, one of which simply says: "required for the effective conduct of war." During World War II, rubber was of strategic importance because it was used in tires, critical to the production and use of vehicles dependent on rubber tires, which in turn were critical to movement of troops and material in the conduct of the war. Today, certain materials used in the production and distribution of electronic circuits that get assembled into equipment used in communications networks are considered strategic by the US Department of Defense, so much so that it expends significant sums of money to maintain alternative sources of supply. Some organizations consider their communications networks strategic and arrange for access to and use of circuits, facilities, and services from more than one carrier.

Making a long-range strategic plan for an organization has its value, but if the plan is flawed or poorly implemented, the effect can be devastating. Long-range strategic plans should be made with care and deliberation. They should also be considered separately from and dovetailed with the annual business plan. Ideally, the long-range plan should be created, or if it exists, reviewed just before the start of the annual business plan and operating budget. One should be done while the other is relatively static. This view is based on the simple fact that when most organizations are created, they are created with a mission, purpose, and some objectives. Very few organizations are created with a limited life. If the organization is a business and is contemplating use of borrowed or equity capital, its founders almost surely prepared multiple-year financial statements and elaborated on what they planned to do to earn a return on invested capital. If a plan is created properly and thoroughly, most of its value comes from the process of creating the plan. Once it's created, it becomes a guide and its execution is subject to change from outside forces, as well as organizational behavior. Many organizations think of their long-range plan as a living document subject to periodic review and change as the entity evolves and conditions around it—over which it has no control—change.

There's no lack of literature on the subject of strategic planning. Strategic planning has gone in and out of vogue as a subject of interest since the 1960s. The basics have to do with the organization or enterprise posing a set of questions and then making a conscious decision to formulate answers, eventually resulting in a so-called long-range or strategic plan. Questions along the following lines should be asked and addressed in detail:

- What is our mission?
- Who are we, and why are we here?
- What is our planning horizon?
- What are all the possible things we could do if we had unlimited resources?
- Which possibilities do we choose, and how do they fit in our time frame?
- How should we go about pursuing our objectives?
- How much can we afford, and are willing, to spend?

- What are the risks and estimated rewards of each project or program we might choose?
- What kind of environment can we expect to operate in during the plan period?
- What regulatory and/or technology limitations or enablers are likely to impact the organization during the plan period?

Many non-profit and government organizations make their long-range plans public. Many of them include great detail on their plans for investment in communications and computer technology and set expectations about the return from their investment. Two very good examples include Fairfax County (Virginia) and Irving Independent School District (Irving Texas).

Private and publicly held commercial businesses almost never disclose anything about the details of their long-range plans. The main concern is preventing disclosure to competitors; however, in the case of publicly held companies, disclosure of any information, short- or long-term, may be deemed forward-looking and must be in accord with applicable laws and regulations. Examples of commercial business long-range plans are difficult to come by. However, Harvard Business School, The Wharton School, and other well-known business schools use the case study method whereby actual cases are disguised and used in classroom lectures and study group activities.

Three good basic references include "Shirt-sleeve approach to long range plans," "Strategic planning in diversified companies," and "The reality gap in strategic planning," which all *Harvard Business Review* articles. The first fits small- and medium-size companies; the second addresses large, multi-divisional companies, and the third is applicable to all organizations because it addresses the critical aftermath of creating a satisfactory plan where planning and reality meet.

LONG-RANGE PLAN CONTENTS

A long-range plan can be a simple document, or it can be extended into a voluminous tome. To a certain extent, the size of the planning document will reflect the size of the organization or corporate entity. The best long-range plans are built much like annual business plans.

Top management or organizational leadership provides a set of assumptions, objectives, and goals. Key staff and lower level management then consider their department's role and responsibilities and make an estimate of the resources required to achieve the goals and objectives within the constraints inherent in the assumptions. The planning cycle need be no longer or more complicated than the annual business planning cycle. Ideally, the long-range plan will be an extension of the annual business plan, with focus on growth targets and performance against those targets each year instead of each quarter in the current year.

At the small- and medium-size business level, most entities are by nature limited to a few products. As an organization grows, the number of products and markets addressed with those products grow. Ultimately, most enterprises operating in the profit-making sector grow to a point where they diversify. The first level of diversification is an additional product in the same or adjacent market segments, followed by new or enhanced products that address new markets. Ultimately, growing businesses are acquired or acquire other companies. As acquisitions are made, the acquiring company is said to operate more than one line of business, while the acquired company becomes a line of business or product line in a larger entity.

Balancing investment and resources in an enterprise where the product or service portfolio is limited to a few specific offerings in a well-defined market segment is one kettle of fish. It involves a two-level structure where organizational leaders decide on a basic set of products and services and define the market in terms of applications and customer set, followed by detailed market research and competitive analysis. This set of intelligence is then used to make a strategic operating and financial plan encompassing the current plan, or next year's annual plan and projections covering 3, and preferably 5, years into the future.

On the other side of the spectrum lies the organization with two or more operating units or divisions, each with its own sales force, product, and/or service set, operating independently of other divisions in a unique market, keeping a unique set of accounting records while using a unique set of human and capital resources. This situation is vastly more complex because it operates on two levels and involves complex trade-offs across multiple divisions or operating units.

The eternal pursuit of the Holy Grail of growth is a continuous consideration of internal or external growth. Do we grow by acquisition, or by internal design and development? This is one of the most often asked questions when considering long-range plans. Another common question: Where do we get the money—are we generating sufficient cash, should we issue more common stock, and is borrowing a viable option? Resolving these issues is the stuff of creating and updating long-range plans.

Non-profit sector organizations and government enterprises don't necessarily use the same terms, but must recognize their sources of income and typically use the same accounting terms such as revenue, cost, assets, liabilities, etc. Sometimes these organizations merge, but less often than profit sector enterprises.

Most of the effort and focus in creating or updating a long-range plan is likely to go into planning and detailing the product or service mix and engineering design and development programs. A successful business depends on a well-defined and managed product or set of products. A well-defined product set comes from continuous study of customer needs, competition, emerging core technology, and evolving standards. Understanding timing and economic parameters that constrain the organization is key to successful, long-term planning. Basic timing associated with day-to-day financial aspects such as the time from receipt of an order to fulfillment of the order varies from industry to industry and entity to entity. But billing to cash collection is about the same for everyone. If the organization designs, makes, and sells products, it is highly likely they build products for inventory, and production time is a key consideration in the operations cycle.

PRO-FORMA FINANCIAL STATEMENTS

A key part of the long-range plan is the financial summary—balance sheet, profit and loss, and funds position statements for the current-year forecast and 5 years into the future. The details in these summaries provide insight and information about the sources of revenue and gross margin in the product and/or service mix, cost to maintain, design, and develop new products, and capital investment. If there is significant debt in the balance sheet, then there may be additional detail showing the amount and payment schedule for each bond or loan. Examples of each of these statements, sometimes labeled *pro-forma*, meaning not real, futuristic, or representative of what will or is hoped will be, as opposed to actual, or historic.

The examples appearing here are representative of an ongoing enterprise typical of many that have either gone public or may be about to go public (i.e., sell equity in the business to a large number of owners or shareholders through a public offering of common stock). Details about the structure of the stock—number of shares and their ownership—are intentionally left out because it's not germane to the subjects at hand, which are revenue, gross margin, net income, product and service design, development, and capital investment. A complete explanation and tutorial on financial statements is beyond the scope of this book; however, consider this an introduction. If you want more insight into the world of financial statements check out Professor Robert West's homepage at: www.homepage.villanova.edu/robert.west. For still more enlightenment and a handy desk reference, see *Finance and Accounting for Non-Financial Managers* by Stephen Finkler. For now, here's what you need to know to get started: The income statement is the key operational tool on a month-to-month basis. It is a measure of the financial performance of the organization over a given time, say a month, quarter, 6 months, three quarters, or a full year. The balance sheet is a snapshot of the financial status or health of the business at a point in time—more specifically, the end of the period covered in the income statement. The statement of changes in funds position tells what happened—where all the money came from, and what it was spent for—during the period of time between the current and previous balance sheet.

Profit and Loss

Table 13-1 is an example of a profit and loss statement for our mythical profit-making enterprise.

TABLE 13-1

Income statement	Forecast current year	Long-range-plan				
(Millions)		Year 1	Year 2	Year 3	Year 4	Year 5
Revenue:						
Product or Service A	125.7	123.8	117.5	117.8	116.7	116.0
Product or Service B	67.5	78.3	99.7	118.0	124.7	132.8
Product or Service C	36.8	65.7	89.4	104.5	116.3	114.0
New Product X	–	1.6	8.5	15.6	42.7	92.5
New Service Y	–	–	–	3.9	14.9	21.6
New Product or Service Z	–	–	–	2.7	8.4	19.0
Total Revenue	229.9	269.5	315.1	362.4	423.6	495.8
Cost of Goods or Services Sold:						
Product or Service A	62.9	56.9	57.9	58.9	59.0	59.1
Product or Service B	32.4	37.6	47.9	57.6	60.1	61.4
Product or Service C	16.5	29.6	39.2	49.6	56.3	55.0
New Product X	–	0.8	4.7	7.3	19.5	46.3
New Service Y	–	–	–	1.8	6.5	10.5
New Product or Service Z	–	–	–	1.3	4.2	9.2
Total Product and Services Cost	111.8	124.9	149.7	176.5	205.5	241.5
Product Design and Support:						
General Engineering & Development	2.3	2.1	2.4	2.6	2.8	3.0
Sustaining Engineering	3.9	3.5	3.4	3.5	3.5	3.6
New Product X	7.6	7.6	2.0	–	–	–
New Service Y	6.3	6.3	6.4	5.8	–	–
New Product or Service Z	4.9	5.7	5.8	3.8	–	–
Unallocated Resource	–	–	5.8	10.1	19.0	21.0
Total Product Design and Support	25.1	25.2	25.8	25.8	25.3	27.6
Gross Margin:	93.0	119.4	139.7	160.2	192.8	226.7
Percent To Sales	40%	44%	44%	44%	46%	46%
SG&A	51.3	53.6	55.8	58.6	64.3	67.5
Interest	0.3	0.3	0.3	0.3	0.3	0.3
Depreciation	1.5	3.2	5.1	7.2	6.5	3.8
Other Cost/(Income)	0.1	0.1	0.1	0.1	0.1	0.1
Sub-Total	53.2	57.2	61.3	66.2	71.2	71.8
Pre-Tax Profit	39.9	62.2	78.4	94.0	121.6	154.9
Provision for Taxes	(19.1)	(29.9)	(37.6)	(45.1)	(58.3)	(74.4)
Net Income:	20.7	32.4	40.8	48.9	63.2	80.6
Percent To Sales	9%	12%	13%	13%	15%	16%

We want to focus on revenue, engineering, capital investment, and depreciation. Capital investment is missing entirely, and depreciation is only a single line. The reason and purpose of showing this statement this way is to foster an understanding of the vague, gray area between public and internal only reporting. This gap is vast and not well understood for the most part, except to those in top management and accounting, where most of the detail is created and kept under tight control. Broadcast and communications engineering and operations professionals need to understand the basic concepts embodied in financial statements and the interaction between revenue, gross margin, product and service cost, capital investment, and depreciation.

The long-term growth and prosperity of the business rests on three factors: revenue and gross margin from current products and/or services, design and development of new or replacement products and services, and capital investment. Organizational sustenance requires that revenues meet, and preferably exceed, all cost of doing business. After all costs are covered, any debt must be serviced in the form of interest and payment of principle. What's left is called income before taxes. However, only interest paid can be deducted before subjecting the income to taxes.

Non-profit and governmental enterprises operate with the same basic requirement except they are not by law allowed to make a profit. Non-profit organizations operate under a specific section of the tax law in most countries. They obtain an exemption certificate initially and must file regular reports of income and cost to retain the certificate. Governmental organizations—federal, state, municipal, and local government bodies as well as organizations they sponsor—are not generally required to obtain tax exemption certificates; however, they generally have reporting and financial accounting requirements imposed on them by the parent or body that created them in the first place. Regardless of their tax or commercial status, these organizations are required to keep accounting records that keep track of the money they receive and pay out. Maybe there's a vague distinction between profit and non-profit by choice and use of the terms income statement and profit and loss (P&L). Both have the famous bottom line, and the bottom line is the fact that survival is dependent on sufficient money coming in to match, or in the case of the profit oriented business, a margin to pay dividends and grow the business strategically.

PRODUCT—SERVICE DESIGN AND DEVELOPMENT

One of the first things that should be apparent about the multi-year P&L compared to the yearly P&L is the time frame. In the instant case, a 5-year planning horizon was decided on. The basic factor driving this selection is the product life cycle of the organization's products and services. In this case, the design and development cycle from start to product introduction is between 2 and 3 years. In other words, it typically takes our mythical company that length of time to organize and perform all the tasks required to get from nothing to revenue producing product or service. Once product and service delivery begins, the product life averages 5 to 7 years. While the 5- to 7-year cycle might be representative of many businesses, recognize that there are many others with significantly shorter, as well as others with longer time frames. Understanding and recognizing the reality of the life cycle, it is no surprise many companies go the acquisition route instead of internal product or service development. After all, why risk losing an opportunity to buy an ongoing, successful business now instead of waiting 3, 5, or 7 years for engineering, marketing, and sales to make one?

Table 13-2 is an example of the magnitude and timing between development cost and revenue.

In practice, revenue is not what we are really looking for. What we're looking for is gross margin—the difference between cost of goods or service and revenue. Typically, gross margin averages 50% to 55% over the life of a product. Each development project is evaluated by a simple return-on-investment measure, where the total development cost is measured as a percentage of gross margin returned over the life of the product. So, in the case of our mythical product X, the development cost is approximately 20% of the projected gross margin return. This relationship can vary widely of course because this is literally one of the most creative parts of managing and running a business. The total return can depend on sales and marketing creativity once the product is ready for market.

TABLE 13-2

Revenue					
New Product X	1.6	8.5	15.6	42.7	92.5
New Service Y			3.9	14.9	21.6
New Product or Service Z			2.7	8.4	19.0
Design & Development					
New Product X	2.4	3.6	1.2		
New Service Y	3.3	3.3	2.4		
New Product or Service Z		3.7	4.8	3.7	

The broadcast engineering or operations professional, or whoever may have just inherited responsibility for dealing with Internet and Telecom likely knows, or has dealt with, the program content development cycle. The basic issues are simply finding out what the programming people want to attempt, and then determining whether or not it can be done with existing resources, or if not, what will be required to do what they want to do and how much capital and operating expense has to be contended with. In turn, programming makes an estimate of how much money can be made with the result and either gets on with it or defers to top management for approval to proceed.

When dealing with Internet and Telecom, there are two perspectives to keep in mind. One, and perhaps the most important, is to make sure the organizational functions requiring communications have access to the best advice and counsel possible with respect to the selection and use of the communications technology and commercial products required by their function. Two, the communications management function is responsible for consolidating and coordinating the requirements into some kind of infrastructure optimized for the entire organization.

These two perspectives require almost constant surveillance of day-to-day current operating costs and project management required to support development and execution of the current annual business plan. The longer term requires knowledge of emerging standards, technology, and commercial off-the-shelf products that can be leveraged into the infrastructure on a longer-range, multi-year basis.

Examining the cost items in the P&L, you may wonder where the departmental expense comes in to play. It all depends on which department. For example, the SG&A item is highly summarized and typically contains several departmental expenses—sales, marketing, accounting, corporate staff, and others of similar nature. The acronym means *selling, general, and administrative expense.* Another example is product design and support. This category likely includes all organizational functions involved in creating products and supporting the customers who buy them. Depending on the size of the enterprise, there may be several departments, each with its own departmental budget, spending ability, and monthly report.

Another place in the P&L where departmental expenses are likely to show up are in the *cost of goods and services sold* category. Product cost —service cost as well, but less likely—includes three components: direct labor, material, and overhead. For example, let's say our mythical company designs and manufactures widgets. And let's say that the company has several types of widgets and maybe even a widget system or two that are sold as individual items to many customers and shows up in the P&L revenue and gross margin in the Product A, B, and C categories.

When accounting set up the system to keep track of all the money involved, they established accounts to capture the labor content and materials used to make the widgets. The manufacturing or production department would have its departmental expense budget, and include management and administrative heads as well as assembly and test or inspection heads. Management and administrative heads are a little tricky in that the expense associated with them may have to be allocated across several products. Other expenses, such as floor space, power, and other utilities, may be allocated as well, based on each product's percentage of the total.

INVESTED CAPITAL

There are two types of capital employed in most any kind of organization: working capital and investment in fixed assets, such as land, buildings, equipment, and other resources required to support the operation. Working capital includes items such as payroll, prepaid expenses, and inventory. Investment in fixed assets is typically considered once a year during the budget planning cycle. Some organizations use a process whereby the capital budgeting process is considered on a project-by-project basis instead of within the context of a strategic plan. Even if the project is a multi-year project, once approved by management and/or board of directors, it goes forward within the context of the organization's project or program management policy.

Capital investment and changes in working capital show up in the statement of changes in financial position, or sources and uses of funds. It's likely that communications capital investment may be planned and made as part of the IT/MIS or broadcast technical operations systems. IT/MIS and communications equipment, networks, and systems require similar focus and effort based on industry peer organizations, technology, and evolving standards. All three functions are cost centers and don't generate revenue and gross margin. But managing operating expense and capital investment have a direct impact on the organization's break-even point, or net income.

In a business where the output is technology-based products and/or services, product planners and managers work hand-in-hand with design engineers and engineering managers to conceive, design, and deliver products to their customer base. The financial focus is on investing money in design and development of a product that will sell at a particular price that can be made for a cost that results in a margin of profit when sold. Considering the fact that products require time to conceive, design, test, and begin manufacturing, revenue and gross margin are not realized until after time and resources have been invested. So the focus is on the amount of gross margin any given product will produce over its lifetime.

Service businesses have the same constraints and issues, except the funds tend to be capital investment instead of design and development expense. For example, in the media industry, newspapers and magazines require printing plant infrastructure. If they don't own the facility, they must outsource the production process and pay the cost plus a profit to the owner. Broadcasters and production houses require significant investment in plant and equipment, but require very little design and development. Communications network service providers are similar to broadcast and other media industry players in the sense that they require very little design and development; however, like the broadcast business, it is very capital-intensive.

So if we have now figured out that businesses in the business of manufacturing and providing services focus their efforts on designing and developing products and services, and the result they are looking for is revenue and gross margin, how does that model play in the IT/MIS and communications spheres where there are no revenue and gross margin? The answer in simple terms: investment and cost reduction. The return on investment yardstick measures the amount of money saved (cost reduction) compared to the investment or cost of capital used to acquire fixed assets. The cost reduction realized may be something as simple as an alternative way to accomplish the same operational function, or it may be more strategic and replace several operational functions. Remember, the bottom line is cost reduction.

The other acceptable strategy for acquisition of fixed assets is increased production capacity, which increases revenue and gross margin. You may hear the term *incremental* revenue and gross margin, which refers to the fact that the entire gross margin falls to the bottom line or increases profit before taxes. This is true only if there is no increase in selling, general, and administrative expense. A kind of mixed-case example in recent years is order entry capability added to a website whereby customers enter orders directly into the company's computer system. Taken alone, order entry isn't of much value, but properly integrated into the IT infrastructure, there's hardly a function in the normal order process that isn't made obsolete. No sales representative, no order entry clerk, no invoice preparation, and if the system can accept and validate credit card numbers or other payment methods, the only thing left to do is ship the product and gaze at the money.

CAPITAL PLANNING: WHERE'S THE MONEY?

Capital planning is basically a decision-making process about how to commit relatively large sums of money for long periods. The money may be unencumbered equity funds or borrowed money.

The money is exchanged for tangible goods and directly related services. Once the asset is created and placed into service, it becomes a revenue-producing asset. Planning and executing capital projects encounter technical and time-related risks. Planning the use of capital involves describing and defining what to do and how much to spend against an expected future return of revenue and gross margin in quantitative terms.

Capital investment, or the results of it, show the use of funds, statements in the depreciation and capital investment lines in Table 13-3. Depreciation is a source of cash funds and capital investment is a use of funds. Depreciation is a positive number, indicating a source of funds, while capital investment is shown in parentheses, telling us it is a use of funds, and subtracted from available funds. Depreciation is a calculation based on value and lifetime of the capital investment. Buried within these two single lines is a highly summarized version of capital investment, including plant, equipment, land buildings, software, hardware, and other fixed assets made during the course of a year. This is the way the information is typically made public, after it is spent. Table 13-3 shows both lines from our financial statement.

Depreciation is generally not well understood by non-financial people; however, it isn't complicated conceptually or mathematically. Briefly, here's how it works: During the planning stage of a project, accounting makes a preliminary judgment about the life of the asset. This assessment determines the number of annual accounting periods over which the asset will be written off or depreciated. For example, equipment may be written off over 5 or 7 years, while software may be written off in 3 years or even expensed outright. As construction or implementation proceeds, there comes a time when the asset is placed into use. When this occurs, an expense amount equal to the value of the asset for that period is transferred

TABLE 13-3

	Current year	Yr-1	Yr-2	Yr-3	Yr-4	Yr-5
Depreciation	1.5	3.2	5.1	7.2	6.5	3.8
Capital Investment	(4.8)	(6.3)	(6.8)	(7.4)	(7.6)	(7.7)

from the depreciation account to the cost section of the P&L. It becomes an expense item, reducing gross margin available as income before taxes. In other words, as long as there is sufficient revenue and gross margin, depreciation reduces the amount of income taxes paid.

Mathematically, here's what happens by example: Let's say that an item cost $5000 and another $1000 to install and make ready for use. The total of $6000 is set up in an account whereby each month, for 120 months (the life of the asset), $100 is transferred from the asset account to the appropriate expense account. The entity's assets are reduced by $100, the monthly expense is charged, reducing gross margin subject to taxes.

COMMUNICATIONS CAPITAL PLAN

Remember that we are concerned most of all about what it takes to move content around. Essentially this involves the basic transport channel and supporting communications. In other words, moving the content requires some amount of bandwidth on a facility. Usually that facility is one-way, except for some type of monitoring or verification that the content reached the desired destination. Beyond this basic, there's consideration of the necessary support functions in the catch-all voice, data, and video, whereby people and computers talk and control what goes on. Essentially, the process is in a few parts. First, define or receive from management a set of strategic actions or directions to consider over some period. Second, estimate the cost and return over the life of each strategy, and finally compare and prioritize the alternatives to reach a point where the most desirable and affordable strategies are adopted and included in the overall capital plan.

Alternative Strategies and Directions

When considering alternative strategies, there may be situations whereby capital investment and projects are mandated by regulatory forces. For example, the Federal Communications Commission says the station must put a digital channel on the air by a certain date or risk losing its analog license and business franchise by another date.

So-called mandated investments come first. Second highest priorities are survival investments. For example, all the other stations in town have converted from using tape and linear editing in news to non-linear editing and file servers. Breaking news stories are consistently running 30 to 45 minutes behind the competition, reducing audience size and ratings while advertisers desert the station. Last are cost reduction and expansion projects, usually nice to do because the funds may be required to pay down debt, or acquire another station or group of stations.

The best of all possible worlds is a situation where the business is highly successful and money is available to consider technological-based change over a reasonable period. Depending on the size, geographic, and market coverage, there can be good logical, sound reasons to consider emerging standards, technology, and products over a 5-year planning horizon.

Regardless, it's likely the results of the effort will be summarized and considered with some being approved while others are placed in limbo to wait for a crisis to arise, or appear in a future planning session.

TABLE 13-4

Balance Sheet	Forecast current year	Long range plan				
(Millions)		Year 1	Year 2	Year 3	Year 4	Year 5
Assets:						
Cash and Equivalent	33.5	38.5	32.0	55.9	34.0	88.8
Trade Receivables	47.5	57.8	68.7	78.9	91.0	107.0
Inventory	38.3	44.9	52.5	59.4	68.8	81.2
Other	1.5	1.5	1.5	1.5	1.5	1.5
Sub-Total Current Assets	120.8	141.4	153.3	194.2	193.9	277.1
Land, Buildings, and Equipment	11.6	11.6	11.6	11.6	11.6	11.6
Other	1.5	1.5	1.5	1.5	1.5	1.5
Sub-Total Fixed Assets	13.1	13.1	13.1	13.1	13.1	13.1
Total Assets:	133.9	154.5	166.4	207.3	207.0	290.2
Liabilities:						
Accounts Payable	26.8	29.2	36.3	41.3	47.5	56.3
Accrued Interest	0.1	0.1	0.1	0.1	0.1	0.1
Other Liabilities	0.0	0.0	0.0	0.0	0.0	0.0
Sub-Total Current Liabilities:	26.9	29.4	36.5	41.5	47.6	56.5
Long Term Debt	3.5	3.5	3.5	3.5	3.5	3.5
Deferred Taxes	0.2	0.2	0.2	0.2	0.2	0.2
Total Liabilities	30.7	33.1	40.2	45.2	51.4	60.2
Equity:	101.8	121.3	126.1	162.1	155.6	230.0
Total Liabilities and Equity	132.4	154.5	166.4	207.3	207.0	290.2
Book Value (Assets − Liabilities + Equity)	203.5	242.7	252.3	324.2	311.2	459.9
Income statement	Forecast current year	Long range plan				
(Millions)		Year 1	Year 2	Year 3	Year 4	Year 5
Revenue:						
Product or Service A	125.7	123.8	117.5	117.8	116.7	116.0
Product or Service B	67.5	78.3	99.7	118.0	124.7	132.8
Product or Service C	36.8	65.7	89.4	104.5	116.3	114.0
New Product X	–	1.6	8.5	15.6	42.7	92.5
New Service Y	–	–	–	3.9	14.9	21.6
New Product or Service Z	–	–	–	2.7	8.4	19.0
Total Revenue	229.9	269.5	315.1	362.4	423.6	495.8
Cost of Goods or Services Sold:						
Product or Service A	62.9	56.9	57.9	58.9	59.0	59.1
Product or Service B	32.4	37.6	47.9	57.6	60.1	61.4
Product or Service C	16.5	29.6	39.2	49.6	56.3	55.0

(continues)

TABLE 13-4 (*continued*)

New Product X	–	0.8	4.7	7.3	19.5	46.3
New Service Y	–	–	–	1.8	6.5	10.5
New Product or Service Z	–	–	–	1.3	4.2	9.2
Total Product and Services Cost	111.8	124.9	149.7	176.5	205.5	241.5
Product Design and Support:						
General Engineering and Development	2.3	2.1	2.4	2.6	2.8	3.0
Sustaining Engineering	3.9	3.5	3.4	3.5	3.5	3.6
New Product X	2.4	3.6	1.2	–	–	–
New Service Y	3.3	3.3	2.4	–	–	–
New Product or Service Z	–	3.7	4.8	3.7	–	–
Unallocated Resource	1.9	–	5.8	13.1	19.0	21.0
Total Product Design and Support	13.9	16.2	20.0	22.9	25.3	27.6
Gross Margin:	104.2	128.4	145.5	163.1	192.8	226.7
Percent To Sales	45%	48%	46%	45%	46%	46%
SG&A	51.3	53.6	55.8	58.6	64.3	67.5
Interest	0.3	0.3	0.3	0.3	0.3	0.3
Depreciation	1.5	3.2	5.1	7.2	6.5	3.8
Other Cost/(Income)	0.1	0.1	0.1	0.1	0.1	0.1
Sub-Total	53.2	57.2	61.3	66.2	71.2	71.8
Pre-Tax Profit	51.1	71.2	84.2	96.9	121.6	154.9
Provision for Taxes	(24.5)	(34.2)	(40.4)	(46.5)	(58.3)	(74.4)
Net Income:	26.5	37.0	43.8	50.4	63.2	80.6
Percent To Sales	12%	14%	14%	14%	15%	16%
Source and Use of Funds						
Sources of Funds:						
Net Income Before Dividends	26.5	37.0	43.8	50.4	63.2	80.6
Depreciation	1.5	3.2	5.1	7.2	6.5	3.8
Working Capital (Inc)/Dec	–	(19.6)	(4.8)	(36.0)	6.5	(74.4)
Cash Provided (Req)/Prov By Ops	28.0	20.7	44.1	21.6	76.2	10.0
Use of Funds:						
Capital Investment	(4.8)	(6.3)	(6.8)	(7.4)	(7.6)	(7.7)
Dividends Paid	(5.0)	(5.0)	(5.0)	(5.0)	(5.0)	(5.0)
Ending Cash Balance	18.2	32.0	55.9	34.0	88.8	22.7

14

PRACTICES, TOOLS, AND TECHNIQUES FOR LONG-TERM SURVIVAL

How often do you hear "He's a survivor" or "She's a survivor?" Usually the implication, or perhaps wishful thought, of the observer is more of a snide remark about the survivor's perceived political connections and less of a true indication of their professional abilities. Setting aside the political aspects and focusing on professional abilities, just what is meant by *survival* and what are the basic ingredients?

First of all, survival as used in this chapter means continuing to exist, grow, and prosper professionally in the context of changing circumstances and surroundings. Well-behaved people who understand the importance of and practice a balanced personal life also find surviving at work far less stressful than those who don't. In this chapter we'll stick to the professional side. In a nutshell, surviving long-term requires interest and effort inside and outside of one's organization.

Inside the organization, it's relationships and responses that count. Being or becoming a source of knowledge about commercially available communications products and services that can enable or help a department manager to meet internal business goals and outside competitive pressures is valuable beyond the normal day-to-day voice and data network support. Understanding the strategic implications of a merger or acquisition, regardless of whether your organization is the target or takeover, can help you be in the right place at the right time, and more importantly, doing the right things. If things don't work to your satisfaction and it becomes time to move on, you are better prepared to initiate and manage personal change if you practice what it takes to stay up-to-date in your field as well as general business practice.

Outside the organization and your daily work environment, there are more places and ways to ensure survival than there is time and money to leverage. Choosing from the available possibilities is the challenge that must continually be considered and addressed. Activities such as conferences, seminars, standards development, and reading can keep you abreast of, and provide insight into, regulatory, technological, and general trends and directions. Depending on your bent for building, another useful activity is to design and make a model of something. Hardware, software, or otherwise, anything where understanding makes its way into the brain via eyeballs and fingers while giving the ears a rest can provide satisfaction, improve self-confidence, and enable clear thinking.

Long-term survival is really a matter of being ready or being in a position to get ready quickly and move onto the next chapter in one's professional life.

CONFERENCES AND SEMINARS

If your background is mostly broadcasting, it is likely you know about and attend the International Broadcast Convention (IBC) (http://www.ibc.org/), the National Association of Broadcasters (NAB) (http://www.nab.org), or both. In addition you probably know of the Society of Motion Picture and Television Engineers (SMPTE), and the Institute of Electrical an Electronic Engineers (IEEE) (http:www.ieee.org). The Broadcast Technical Society technology-based conferences and training courses. For example,

SMPTE conducts a seminar in the fall, usually in November, and a second one in February each year. These conferences are technical in nature, and usually include subject matter indicative of trends and directions from emerging core technology to new products and operational methods. SMPTE also sponsors four quarterly television standards meetings per year, plus others concerned with motion picture standards. Most regional SMPTE chapters sponsor regular monthly meetings where guest speakers are invited to present material on subjects of interest in the local area.

The IEEE Broadcast Technical Society conducts two symposiums a year: One in the fall and a second during the NAB conference in the spring. These conferences attract some articles and seminar subject-matter experts specifically recruited by society leadership based on member interests. Continuing education credit is usually granted those registering and attending.

On the communications side, there are similar level conferences and conventions you will find helpful in building and maintaining knowledge of the Internet and Telecom fields. Examples include:

Supercomm—Similar content structure as NAB and IBC; 500+ exhibitors in 2003; www.supercomm2003.com/
Optical Fiber Communications Conference: Oriented at communications on fiber media and related interests, 500+ exhibitors in 2003; www.ofcconference.org/
Updated listings can be found at:
dir.yahoo.com/Business_and_Economy/Business_to_Business/ Communications_and_Networking/Telecommunications/ Conferences_and_Trade_Shows/
Internet: The Internet Society sponsors or sanctions many conferences; www.isoc.org/isoc/conferences/inet/
Computing: Siggraph—one of the better computing conferences with good participation by people and companies interested in generation and display of images and graphics; www.siggraph.org/ s2003/
Association of Computing Machinery (ACM): Several conferences a year oriented a several subsets of computing—grid, mobile, parallel, etc.; www.acm.org/events/

Backing away from the purely technical aspects of seminars, there are numerous non-technical seminars and courses available from local colleges and universities to businesses that are in the business of training and education. One of the most popular subjects is *finance and accounting for non-financial managers*. Search the subject on the web and pick two or three to investigate.

READING MATERIAL

We all have limited time for reading anymore. With the advent of the worldwide web, there are growing almost unlimited numbers of newsletters. And the supply of magazines, free and paid subscription, seems to be growing. Visit one of the major chain bookstores that have sprung up in recent years and it's not difficult to become overloaded. The overall challenge seems to be deciding what, if anything, to acquire.

The bibliography at the end of this book is a cross section of many years' collection. The references are but a slice from literally thousands of pages of articles seen and read over many years. The better ones have a way of sticking around and appearing when it seems most opportune.

WEB RESOURCES

Salter's quote in Chapter 2 about "nobody owning the Internet, and there being no central control, and (therefore) nobody can turn it off" is one of the more profound realizations others come to when considering this valuable resource.

If you don't want to trek to the local bookstore or nearest library, turn to the web. New or used, it's not difficult to find most books if they have been in print in any reasonable quantity in the past few years. Search on the subject you're interested in and it's highly likely there are books, articles, and all manner of printed material. Two of the three articles from *Harvard Business Review* referenced in Chapter 13 were easily found on the web. All three were found in reference material in other papers and lecture notes.

The one caution is due diligence. Don't just pick out the first or what seems like the most knowledgeable material. Study several and see if the same message or messages come through each time. Then check it against your own reasoning skills.

TECHNOLOGY DRIVERS

Perhaps the most fruitful source of knowledge comes from those in academia and industry who continuously develop core technologies. One view of the value chain has core technology as a key ingredient in hardware, software, and equipment. Services simply cannot exist without the latter. And the best of equipment, software, and systems wouldn't be the best without key core technologies. One only has to look at the real world to see that it is continually changing and evolving on a natural pace. What may seem chaotic and unpredictable on the surface teems with structure underneath. Countless examples, such as time division multiplexing (TDM) or plesiochronous digital hierarchy (PDH), synchronous optical network/synchronous digital hierarchy (SONET/SDH) followed by Internet protocol (IP)/SONET/SDH evolved over many years. It's not difficult to get into that loop and with only a few simple web searches find out that the next generation of equipment and systems are now deeply embattled in pushing standards development in the general form of next generation multi-protocol label switching (MPLS).

The Internet camp calls it generalized MPLS (GMPLS); the ITU side calls it automatic switched optical network. Another example in the same area is resilient packet ring technology maturing as the next generation of optical ring transmission technology under the sponsorship of IEEE 802.17.

To illustrate the characteristics of the nature of how technology evolves into services, economics, and timing, one only has to put a few numbers together to see how this field plays out. Figure 14-1 shows this in graphic terms.

It's not difficult to see that sometimes core technology development and evolving standards run for 3 to 5 years before any tangible level of hardware and software start to take hold and mature. And only

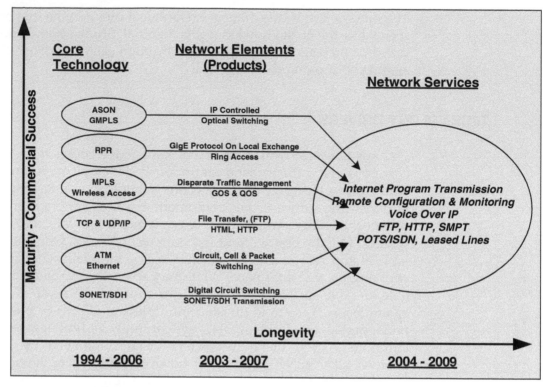

FIGURE 14-1

after many times the investment in real products does a broadcaster or communications business start investing in those products to begin generating revenue and gross margin.

So anyone telling you about technological surprise has been asleep or not paying attention. Once you understand this picture, it becomes a fairly simple exercise to monitor technical papers from the academic community to see where the manufacturers are likely to place their bets in the form of technology development and design, which leads to their press releases about their latest new gadget or discovery and how that might impact their future in our industry.

An interesting example of where fiber development may be headed appeared in the *New York Times* on August 26, 2003 in an article: "New Telecommunication Tools May Emerge From the Deep."

"While in San Francisco for a scientific conference last year, Dr. Joanna Aizenberg, a research scientist at Bell Labs and senior author of the *Nature* paper, wandered through shops looking at shells and other collectibles from the sea.

"In one shop she spotted a Venus' flower basket sponge, a creature with an intricate hollow latticework skeleton that lives thousands of feet deep in the western Pacific Ocean. What caught her eye was a ring of glassy filaments at the sponge's base that once tied it to the ocean floor. She bought the sponge. Back at the laboratory, Dr. Aizenberg and colleagues from Bell Labs, Tel Aviv University, and OFS, a Lucent spin-off, discovered that the filaments, about the length and thickness of hair, also carry light. While other researchers discovered a few years ago light-carrying fibers in a sponge off Antarctica, the fibers of the Venus' flower basket sponge are exceptional because their structure is 'strikingly similar' to those of commercial optical fibers that ferry pulses of light in telecommunication systems, Dr. Aizenberg said. 'Nature came up with exactly the same design millions of years ago,' she said. 'I would be surprised if it's accidental.'"

How many years it will be before this discovery shows up in the field in the form of real fiber is hard to predict and very easy to follow. Just subscribe to the *New York Times* reader alert service on the subject and every time something is published or released, it's likely the subscriber will get an email about it. Such is the new way things are done with Salter's Internet that "nobody owns and nobody can turn off."

STANDARDS COMMITTEE WORK

Another fruitful source of knowledge is standards development committee work. You can simply monitor the work output, or you can join and contribute. The structure of all formal standards work is built around volunteerism and openness. A significant amount of energy is devoted to gaining exposure to the work, to potential users or customers in the form of publishing reports, and sharing the work as consensus is reached. The International Organization for Standardization (ISO) (www.iso.org) and ITU are global standards organizations that have their structure built around delegations from each member state. The American National Standards Institute is the North American umbrella organization whose rules are followed by SMPTE, IEEE, and other standards development organizations in the United States.

As new core technology and processes evolve and attract standardization attention, knowledge about the details of what it does and how it does it become public, in most cases long before it shows up in a commercially available product. Similar behavior is found in the communications world as well as in the broadcast world. Consider for example how many years the Advanced Television Systems Committee has existed and continues to exist for the purpose of developing and maintaining digital television standards.

APPENDIX

PROGRAM CONTENT PAYLOAD SIZE FOR TDM ATM AND IP NETWORK TRANSPORT

Defining content or other payload size is critical to efficient use of network transmission facilities. When program content gets transported in non-real time (i.e. not live or pre-recorded, and not "streaming"), it gets moved by moving the physical media it is stored on, or via a network using file transfer protocol (FTP).

When program content moves in real time, network bandwidth or connection speed is fixed and must be sufficient to accommodate the payload in terms of bits-per-second. When content moves on physical media or via FTP over a network, speed of movement or bandwidth available in the transport channel can be fixed or variable. This variation is intolerable for real-time content transport, but it is a trade-off against the amount of time it takes to deliver the content. Table 1 contains examples of payload bitrates in megabits per second across a range of selected arbitrary compression ratios. These examples are for real time streaming transport only.

Payload may be only the content, or it may include overhead bits to improve end-to-end bits-to-errors ratio (BER) reliability, and/or robustness. The other side of the coin asks: How much bandwidth is available on a given transmission facility?

Content transport on TDM facilities is the simplest and easiest to understand. It is also a reference model for ATM and IP transport media performance measurement and assessment. Figure 1 is a block diagram showing the functions making up multiplexing of audio, video, and data as might be encountered when mapping elementary program content streams into a transport stream and then on to a

TABLE 1

Payload Speed and Content Transfer Rates

High Definition – SMPTE 292

		Real Time Transfer Rate		
Payload Bitrate (Mbs)	Compression Ratio	MB / Second	MB / Minute	MB / Hour
1485.0	None	185.625	11,137.500	668,250.000
560.0	2.65	70.000	4,200.000	252,000.000
360.0	4.13	45.000	2,700.000	162,000.000
270.0	5.50	33.750	2,025.000	121,500.000
155.0	9.58	19.375	1,162.500	69,750.000
100.0	14.85	12.500	750.000	45,000.000
75.0	19.80	9.375	562.500	33,750.000
38.0	39.08	4.750	285.000	17,100.000
19.4	76.63	2.423	145.350	8,721.000

Standard Definition – SMPTE 259

		Ratio Real Time Transfer Rate		
Payload Bitrate (Mbs)	Compression Ratio	MB / Second	MB / Minute	MB / Hour
270.0	None	33.750	2,025.000	121,500.000
67.5	4:1	8.438	506.250	30,375.000
54.0	5:1	6.750	405.000	24,300.000
38.0	7:1	4.750	285.000	17,100.000
27.0	10:1	3.375	202.500	12,150.000
19.4	13.9:1	2.425	145.500	8,730.000
13.5	20:1	1.688	101.250	6,075.000
9.0	30:1	1.125	67.500	4,050.000
7.0	38.57:1	0.875	52.500	3,150.000

transmission channel. The reverse is also shown in which content elementary streams are recovered from the transmission and multiplexing processes. Before explaining how to calculate or estimate payloads for network channels, bit stuffing needs to be explained.

Bit stuffing simply adds bits to a signal to make up for differences in digital network channel hierarchy clocking and timing.

Network clocking and timing parameters are based on time division multiplexing techniques. This means each channel of a given facility runs on a constant clock, resulting in fixed time intervals of equal duration for each unit of information contained in a byte (octet) or sample.

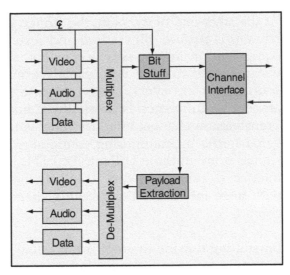

FIGURE 1 Functional Block Diagram Multiplex Channel Interface

Channel boundaries occur at fixed intervals marked by framing or overhead bits. Digital signal 0 (DS0, 64Kbs) resides at the bottom of the hierarchy. Multiple DS0s (24 or 32) created from a common clock and multiplexed with other peer streams of channels to build higher order groups of streams capable of carrying disparate traffic produced by independent standard clocks. The channels run in near synchronous mode but carry asynchronous traffic.

For example, DS3 channel runs at a nominal rate of 44,736,000 bits per second. To keep the channel unique and separate from peer channels, framing bits, also called *overhead,* are required. What is left is called payload. In the case of a DS3, the framing overhead amounts to 526,306 bits per second, which leaves the remainder of 44,209,694 available for payload. (See Table 4-1 for other PDH and Table 4-3 for SONET/SDH channel and payload rates.) However, 672 DS0 (voice channels) running at 64Kbs amounts to only 43,008,000. What about the other 1,201,694 bits? This is where the technique of bit stuffing comes in. These "left-over bits" can be divided up amongst lower order streams to make them fit within a time window compatible with the higher order multiplexed stream. This allows them to wander around, within certain well-defined tolerances, and enables the network accommodate them.

At the other end of the path, the stuffed bits are dropped, and what remains is *payload*. Bit stuffing and reversal may occur in premises equipment, inside the network, or both. Bit stuffing inside the network is used to cope with signals from different multiplexers or sources that are synchronous within a specific range or tolerated amount of difference between their clock and the network clock. Transmission channel interfaces enable the network to carry aggregated traffic by configuring contiguous DS0 or higher rate digital signals from multiple DS0 into DS1 (T1/J1), E1, E3, and DS3 (T3/J3) interfaces. When network transmission channels are configured with these interfaces, the resulting channel is *unchannelized* or *clear channel*.

Bit stuffing outside the network is usually very application specific, such as in a data or video application that may require the equivalent of several DS0s, less than a DS1, DS2, or DS3, as a stand-alone requirement. Bit stuffing might be used to make up any difference between the application's requirement and the fixed, standard network channel is configured to provide.

The concept of a network payload is very apparent in SONET/SDH transmission (see page 98 and 99), in which the payload of 51.8Mbs includes extensible overhead that can accommodate between 1 and 192 51.8Mbs channels. These channels, called *STS1* for *Synchronous Transport Stream 1*, can, in turn, accommodate lower rate channelization to carry legacy PDH payloads. Alternatively, it is possible to map other channel structures capable of carrying program or other content encapsulated in non-PDH structures, such as ATM and IP.

PROGRAM CONTENT TRANSPORT ON TDM CHANNELS

From the earliest days of TDM, practical encoders and decoders have been and continue to be used to transport NTSC and PAL video and analog audio. These devices are designed to produce an output that runs over DS3 or E3. This means they are hard-wired to produce a nominal 30Mbs payload bitrate. In practice, the payload is an *octal number*. In the instant example, the maximum payload possible is 30,720,000 bits per second. To provide some measure of tolerance, or *guard-band* as it's called in RF channels, let's assume the next lower octal number will be our payload rate. Determining this number is a bit of trial and error. Logically, it should be an octal number as well. This means we start with some small number such as 1,048, subtract it from the payload and then divide by 8 to see if the result, 30,718,952, is an octal number. And, yes, that's an octal number. What's important is that the payload fit the channel. It must get transported without errors caused by not enough bandwidth. Incidentally, in this case, it will fit both E3 and DS3.

So what happens to the rest of the channel? It's simple; just stuff it with null bits. As long as null bits don't come in the wrong place, which, in this case, is outside the payload, the network doesn't care. What the network looks for are the framing bits. As long as the framing bits are there and in the right place, the network is fat, dumb, and happy.

FEC may or may not be required, depending on the requirements and what the business proposition can afford to pay. Also, the underlying transmission facility can exhibit drastically different performance. For example, satellite or terrestrial radio channels are far more vulnerable to errors than fiber. Bit error rate in transmission channels is dependent on receiver sensitivity or the ability of the receiver to detect low-level signals, or distinguish between noise and signal. When the RF carrier signal at the receiver antenna terminal is low, the receiver can easily mistake a noise spike or pulse for a digital signal transition. Fiber behaves the same way, except that it has several orders of magnitude greater sensitivity than radio receivers do, plus many times the bandwidth possible on any given transmission path.

In practice, order an E3 or DS3 from your friendly service provider, connect the codecs, and operate the network.

What about ATM and IP framing and protocols? There are as many possibilities as there are protocols. It is instructive and enlightening to compare two more—ATM over PDH/SONET/SDH and IP over PDH/SONET/SDH.

ATM Transmission

The case for ATM transport is more complex than straight TDM but less complicated than IP. To further simplify, assume constant bit rate transport and use of ATM adaptation layer 1 (AAL1) for any number of reasons that area beyond the scope of this Appendix.

Here's the story and an example:

Recall the ATM transport mechanism is a 53-byte cell. 47 or 48 bytes of the 53 are used for payload, and the rest is overhead. The most important thing to recognize and not ignore is the fact that ATM network access on a (PDH) DS3 must fit within the clear channel rate of a DS3 as outlined above (i.e., [44,736,000]-[526,306]=44,209,694bps, which is a brick wall boundary that the network will not allow to be exceeded under any circumstances). This is the maximum potential size ATM payload that can be carried by the TDM DS3, layer 1 network facility. The ATM payload crossing the network channel interface cannot exceed the precise number of 44,209,694 bits each second. Now the fun begins. First, assume we will be using 48/53-byte cell structure. Next, understand and analyze this picture. 53bytes of 8bits equals 424 bits in each cell. There are 44,209,694 bits to put ATM cells into. A little simple arithmetic reveals that the available bandwidth can accommodate 104,268 of these cells or carriers. This quantity requires 44,209,632 bits—no more, no less. The difference of 62 is not 53. It is possible the available bandwidth could carry as many as 104,268 53-byte cells. Yes, but not in every successive second. So what has to happen is organized transfer with some organized bit stuffing. One simple approach could be to transfer 104,269 cells and follow that with the 62 bits of stuffing. In practice, that's just fine because the utilization is almost 100%.

After the ATM channel is fitted to the PDH/SONET/SDH channel, next comes how to fit the program content into the ATM structure. Much like the TDM structure, the ATM structure wants to see 48 bytes of payload that it attaches its overhead to and sends it on its way. At the other end, the reverse happens, whereby the payload is subjected to a reversal process and extracted from its transport structure. Simple arithmetic tells us those 104,268 cells per second, with

424 bits in each, amounts to a maximum content payload potential of 33,783,156 bits per second.

Estimating and configuring an ATM network to transport program content becomes a matter of defining picture quality and sound fidelity in terms of aggregate program stream bandwidth and selecting the size of the ATM channel to accommodate that stream. One or more program streams or payloads can be accommodated on a single E3/T3/J3 access facility as long as the total aggregate of all streams does not exceed the 33.7Mbs maximum.

IP Transmission

IP transport is an entirely different matter. First, early IP networks— the *Internet*—were implemented on ATM backbone networks. This meant that the basic path a set of information traversed included IP encapsulation, mapping to ATM, and then TDM, which likely included SONET/SDH as the underlying layer 1 transport.

More recently, as traffic has grown and technology has matured into more capable and sophisticated network elements, IP has moved beyond ATM and directly on to SONET/SDH. Without consideration of such things as MPLS and other mechanisms to deal with GOS and QOS issues, it is instructive to consider Transport of the same program content over IP.

A couple of important points to recognize and not forget: (1) IP networks rely on the underlying transport layers—Layers 1 and 2 for clocking and timing. Once a set of information is taken from the source and placed into packets it has no timing relationship until it reaches the other side of the network and the reversal of the encapsulation process takes place. (2) The packets may take different routes through the network and experience varying degrees of delay relative to one another. Obviously, this wouldn't work too well, so there must be some kind of mechanism to ensure the information the packets carry is placed in the correct relationship to each other. How to do this is a very big story, and outside of the scope of this appendix. In short, the answer is protocols—processes that take the program content and encapsulate it into packets that the network carries and delivers, after which it is removed from the packets reassembled into program content.

One example, and there are many, uses IP, UDP, and RTP where the sequence of events or results of running the protocols through their full cycle produces a sequence starting with the headers followed by the content. Picture 20-byte IP, 8-byte UDP, and 12-byte RTP headers and, finally, 4-188 byte MPEG packets with the content in them.

Equivocating this to the TDM and ATM examples above is a bit of a convoluted exercise, but here goes:

- Each repetitive cycle includes 792 bytes (6,336 bits) total
- 752 bytes of content
- 40 bytes overhead (20+8+12)

Note: The above is all static numbers and contains no clocking, only the implications of clocking, conspicuous by it's absence.

After thoroughly understanding this picture, how do we relate payload to a network transport facility? The classic Internet Netizen approach is simply to "give me all you've got." If that were possible, the content might get there before its time came for presentation, and wouldn't that be a can of worms? More to the point, IP-layer 3 transport depends on lower layer 2 for clocking. This says there are bandwidth constraints and structure, so like carrying ATM over PDH or SONET/SDH we can simply take the same E3/T3/J3 channel interface and determine maximum potential payload under the constraints imposed by IP, UDP, and RTP protocols. Now it becomes a simple arithmetic exercise, not unlike that used in TDM and ATM.

First, the maximum potential payload that can be expected to be available from a DS3 channel interface is 44,209,694 bits. This means only so many packets can pass. Since the rate must be constant and each frame consists of 6,336 bits, dividing the available channel rate by 6,336 results in a fractional number (6,977.5). So drop the fraction and subtract 4 making it 6,977 frames, which is an octal number. Now multiply 6,977 times 6,336 bits per frame, and the resulting maximum IP payload is 41,973,632bps.

TABLE 2

Payload Capacity in TDM, ATM and IP PDH Transmission

TDM	Bits	Octal Y/N	Bytes	Circuits
DS3 Channel Rate	44,736,000	Y	5,592,000	
DS3 Framing Overhead	526,306	N	65,788.3	
DS3 Payload Capacity	44,209,694	N	5,526,211.8	
Transmission Unit Payload	64,000	Y	8,000.0	
DS3 Facility Payload	43,008,000	Y	5,376,000.0	672.0
Required	1,201,694.0	N	150,211.8	
Total (Payload+OH+Stuffing)	44,736,000	Y	5,592,000.0	
ATM	**Bits**		**Bytes**	**Cells**
ATM Cell Payload	384	Y	48.0	
ATM Overhead ("Cell Tax")	40	Y	5.0	
Transmission Unit Payload	424	Y	53.0	
ATM Cells On DS3 Payload	44,209,694	N	13,033.5	104,268.1
Next Lowest Octal Number Evenly Divisible by 424	44,209,632	Y	5,526,204.0	
Additional Stuffing Required	62	N	7.8	
Overhead	526,368	Y	65,796.0	
Maximum Practical ATM Payload in DS3	44,209,632	Y	5,526,204.0	104,268.0
Total (Payload+OH+Stuffing)	44,736,000	Y	5,592,000.0	
IP	**Bits**		**Bytes**	**Packets**
IP Payload (4 × 188 Bytes)	6,016	Y	752.0	
IP Header	160	Y	20.0	
UDP Header	64	Y	8.0	
RTP Header	96	Y	12.0	
Sub-Total – Overhead ("Packet Tax")	320	Y	40.0	
Transmission Unit Payload	6,336	Y	792.0	
IP Packets On DS3 Payload	44,209,694	N	5,526,211.8	6,977.5
Next Lowest Octal Number Evently Divisible By 6336	44,206,272	Y	5,525,784.0	
Additional Stuffing Required	3,422	N	427.8	
Overhead	529,728	Y	66,216.0	
Maximum Practical IP Payload in DS3	44,206,272	Y	5,525,784.0	6,977.0
Total (Payload+OH+Stuffing)	44,736,000	Y	5,592,000.0	

Next, what's the maximum program content payload? Easy, just divide the bits into headers and content. Content in each frame

TABLE 3

TDM ATM and IP Transport Comparison

	Channel	Overhead	Available
TDM	44,736,000	526,306	44,209,694
ATM			44,209,632
IP			44,206,272

amounts to 6,016 bits. Multiply that by 6,977, and the resulting content payload is 41,973,632bps.

Summarize the numbers from the three alternatives and this is the payload from a network perspective of the three alternatives—TDM, ATM, and IP transport techniques. Table 2 shows how the three techniques compare when given a fixed rate program content payload. In practice, more detailed design analysis on ATM and IP is likely to result in a set of trade-offs that can be used to structure proposal preparation and analysis with suppliers.

From a content perspective, the payload is not a single entity. The numbers above are summarized from allocation of bits to audio, video, closed captioning, and other data that may be required such as vertical interval test signals. As a first order estimate, this process can be built from basics by assuming or taking rates for each, summarizing them, and adding markup percentages to account for MPEG transport stream overhead and forward error correction to arrive at a program content payload estimate. The markups are summarized as follows:

- MPEG Transport Stream   3%
- FEC   3.125% (On top of MPEG Markup)

GLOSSARY

Analog Analog means "representative of." If something is analogous, then it is a copy or representative of the original. In communications and electronics work, the term *analog* is used to mean a particular type of circuitry or manner of dealing with voltage and current. A good analog signal is absent distortion when compared to the original. A microphone converts sound waves (acoustical energy) into electrical energy—voltage and current. A speaker converts electrical energy into acoustical energy. Distortion of any kind along the path between source and destination is undesirable technically and may not be commercially acceptable to the viewer or listener. Electrically or electronically, an analog waveform is a representation of voltage at any given instant or over a period of time. It can be stable or varying in amplitude. The polarity may be constant or reverse in regular or irregular intervals.

In practical broadcast engineering, *analog* is a label to characterize NTSC, PAL, SECAM video, and non-digitized audio signals. In today's digital world it's important to recognize and remember the following two things: (1) content or the signals representing the content must be digitized or present in digital form to be carried on digital transport media; and (2) video is structured, while its counterpart, audio, is relatively unstructured. Program content—audio and video signals in analog form and digital media mix about as well as oil and water.

Analog-to-digital and digital-to-analog conversion   If a signal is in analog form and needs to be carried on digital media, it must be converted to digital. If the signal is in digital form and must be displayed or heard, it must be converted to analog form. Conversion from one form to the other requires care; otherwise, the process can produce undesirable or unwanted side effects rendering the signal useless for commercial purposes.

Carrier   Carrier is a truncation from classical definition of regulated entities designated *Common Carrier,* meaning the entity is required to offer the same level of service to everyone at the same unit price and under the same terms and conditions on file with a regulatory body, such as a state public utilities commission or the Federal Communication Commission. Also see LEC, ILEC, CLEC, or IXC.

In 1982, AT&T entered into a consent decree with the US Department of Justice by breaking itself up into local and long distance businesses. Carriers were then designated local exchange (LEC) and long distance and Inter-Exchange (IXC). In 1996, legislation was enacted into law, and the local exchange carriers were re-designated into incumbent (ILEC) and competitive (CLEC) categories.

Carrier is also a technical term applied to carrier signals, such as the carrier signal of a radio or television station transmitter. *Subcarriers* are sometimes placed on carrier signals and modulated in similar fashion as carriers.

Occasionally, a T1 transmission channel is referred to as a *T1 Carrier* because the 1.544Mbs digital stream carries 24 discrete channels of 64Kbs each.

CODEC   A truncation of the words en*code* and *dec*ode, sometimes in capital letters, it is also used in communications network elements to describe a device that is full-duplex (i.e., it sends and receives simultaneously). The term is occasionally applied to electrical-to-optical and optical-to-electrical converters used to interface SMPTE 259 or SMPTE 292 television signals to optical fiber transmission facilities.

Convergence   Using a common access and/or transport facility to carry disparate traffic is called *convergence.* Configuring a single network channel so it can carry Voice, Data, Video, and Internet traffic instead of using individual E1/T1 or E3/DS3 channels for each traffic type is more efficient and typically lowers unit cost of bandwidth. However, this results in putting all the eggs in one basket and may not meet availability, reliability, robustness, or any of the three requirements. So-called *Triple Play Networks* are claimed to be capable of carrying voice, data, and video.

Compression and Decompression   Sometimes used interchangeably with *encoding* and *decoding,* respectively, but compression and decompression are entirely different processes. Compression can be either lossy or loss-less. Most compression used today is lossy, meaning that some part of the information from the original signal or file will be missing upon decompression. Choosing the amount and parameters of the compression process should be done within the context of what the signal will be used for and how it will be judged for acceptance or rejection when it is decompressed. An equally important consideration is how many times the signal will undergo lossy compression and decompression before being used (displayed or listened to). See Encoding and Decoding.

DACS   Digital Access and Cross Connect Switch is a network element used mostly by Telecom service providers to groom segments of transmission bandwidth. DACS typically fits between the transmission facility and switching or routing equipment. So-called *Private Lines* are provisioned using DACS to aggregate and partition off sub-T1 (also called *Fractional T1*), T1, E1, E3, or DS3 bandwidth between two circuit endpoints to make up the Private Line. May also be labeled a *Slow Switch*.

Digital   Digital is a common term for either signals or information in digital form or the transport medium—tape, CD, or any network capable of carrying the digital signal. A digital waveform or voltage only has two states—on or off; the signal is either there or it's not there. It's often said to be a 1 or a 0 (i.e., something or nothing). The timing associated with the transition from one state to the other can be synchronized with a clocking signal—*synchronous* or if it's not it's said to be *asynchronous*.

DVB-ASI   Digital Video Broadcasting—Asynchronous Serial Interface is an increasingly popular interface selected by designers for use in digital television facilities. Broadcast engineers with a good understanding of SMPTE-259 like it because it's a lot like what they already know. Communications engineers like it because it's similar to E3, DS3, and STM1. What's to like? BNC Connectors and 8281 coax, for starters. Is this convergence at last?

Facility   A facility is generally, something physical, such as a building with equipment and amenities used to produce content or provide service. Examples include studio facility or transmission facility in the broadcast world. Telecom facilities are often called services. For example, it's quite common, but a misuse of the term to say or write *T1 Service* or *DS3 Service* when describing or meaning a T1 or DS3 access or transport facility.

FEC   Forward Error Correction is Bits or Bytes of information added to content that can be used to detect and reverse errors caused in the transport process. For example, terrestrial and satellite wireless (radio) transmission is said to rely on 3/4 or 7/8 FEC. Take this to mean that for every 3 units of payload, add 1 unit, or for every 7 units, add 1 unit of specifically coded information to be used to detect and correct errors in transmission.

Interface   An interface is a connector, or connection point, sometimes also called *service handoff* or *service demarcation*. Such a designation is most useful when it carries a reference to a single standard such as G.703, RS-232, DVB-ASI, or IEEE 802.11.

Internet   The Internet is a global network based on a set of protocols, depending on units of data encapsulated into packets or protocol data units. Each unit or entity consists of a header followed by a payload. The header contains all the information necessary for routing or moving the payload through the network.

IP   Internet Protocol

IP Address   The IP address is the unique number assigned to a host or other devices in dotted decimal format. It is used to route packets through the network using packet switching techniques. The IP address is functionally similar to telephone numbers used by the PSTN to route phone calls with circuit switching techniques.

IP Multicast   IP multicast is a routing technique used to route packets from a single source or location to multiple destinations using a common address. Functionally similar to classical bridged networks built on analog and TDM networks and point-to-multipoint ATM networks.

IP Unicast IP unicast is the act of sending IP datagrams, or protocol data units, from one location or address to a single destination or address. Conceptually, IP unicast is the same as a point-to-point connection or private line in a TDM network or a private virtual circuit in an ATM network.

IP Broadcast When one or more packets are sent to all hosts or nodes on a network by using a single address, it's referred to as *IP broadcast*. Sometimes it's called a special form of multi-casting. Ethernet protocol found in LANs uses a scheme whereby all communications between devices on a local area network initiate sessions between each other by broadcasting or calling to a single address listened to by all the devices on the network.

Broadcast Address The address reserved for sending the same packets to all stations on a network is the broadcast address.

Broadcast Domain The broadcast domain is a set of devices or network elements that will receive packets from any one of the other devices or elements in the set.

Broadcast Storm Sending or broadcasting to all devices on the network simultaneously, requiring bandwidth beyond the capacity of the network, causing an overload or crash is a broadcast storm.

ISDN Integrated Services Digital Network

Latency or Delay The time (in microseconds or milliseconds) between the absolute time a signal is present at the input terminals and the later absolute time the signal appears at the output terminals of a piece of equipment is called *latency* or *delay*. These terms are more broadly used to indicate a similar measure of the time it takes a signal to transit a network. For example, the transit time of a single satellite path consisting of one uplink, the satellite transponder, and the downlink is typically 250 milliseconds, while a typical terrestrial path between the same two points will be less than 50 milliseconds. Latency or delay becomes an important consideration when these kinds of network links are being planned and implemented to support program production where talent engages in conversation or interviewing.

Media Media is the transmission link that carries the content. Examples include coaxial cable, copper wire, radio waves, waveguide, and fiber. Magnetic and optical storage enable physical transport as an alternative or in addition to network or electronic transport.

Multi-Media   Multi-media (also spelled multimedia) is a truncation of the word *multiple*, appended to the plural of medium, *media*. Meaning different things to different people, its use herein is avoided in an attempt to reduce confusion.

Network   A network is a collection of equipment, software, and systems connected together physically and configured to provide services. Networks are generally thought of in the context of their geographical or physical coverage. Networks may provide one- or two-way services with symmetrical or asymmetrical bandwidth and multiple classes of service. The term, as used in this book, includes electronic communications facilities and specifically excludes physical transportation facilities. Content on media can be read from the media and placed on a network for transport in real time or as a file transfer. Networks are implemented with different conceptual approaches, standards, and technologies.

NID   A Network Interface Device is typically a piece of equipment with standard television and network interfaces on the customer side and network interfaces on the network side. It performs the function of mapping the digital television payload to the network channel and may also be used to aggregate voice and data traffic into and out of the network.

Port   Port is one of those widely used terms meaning the input, output, or both of equipment, network node, or software application.

POTS   Plain Old Telephone Service

PCR   Program Clock Reference (MPEG Jargon)
Primary Clock Reference (Telco Lingo)
Peak Cell Rate (ATM Speak)

Private Line   A privacy line is typically a point-to-point circuit or facility provisioned by a common carrier or service provider according to a formal tariff or on an individual-case basis with specific amount of bandwidth and capabilities not shared with others. Classical Data Communications people use the term *Leased Line*.

Protocol   A set of rules used by devices in a network to exchange information is called a *protocol*.

PSTN   Public Switched Telephone Network

Picture Quality   A phrase of almost universal use as a benchmark is *Broadcast Quality*, a subjective term meaning "it looks and sounds like a TV set." Classical television engineers use a set of metrics including parameters for resolution, color fidelity, and contrast range among others. Focus, though, is mostly on the use of TV lines as an indicator of a component or system's ability to resolve detail in an image. To a broadcast engineer, the term *resolution* (see below) translates to TV lines. Resolution in print and film means "line pairs per millimeter." Outside the television world, resolution is used to mean pixel count. Another similar term found in monitor specifications is *Dot Pitch*.

A properly designed and configured content transport network, connection, or facility will not impact picture quality so long as it is operating within bandwidth and error rate performance specification parameters.

Resolution   Resolution is determined by the resolving power of the pickup and display devices and other system components in between. Spatial resolution is a measure of the ability of a component or system to resolve detail in an image. Temporal resolution is the image repetition rate in continuous motion capture and display. Television displays use FPS—frames per second—and computer displays use *Refresh Rate*.

Resolution—also   With the promise of convergence on the imaging side (as opposed to converging disparate traffic types on a network access facility), one should be aware that the motion picture industry uses the term line pair per millimeter (LPPM) and the print/graphics industry uses dots per inch (DPI). And it bears repeating again so you won't needlessly traverse blind alleys: *A properly designed and configured content transport facility, connection, or network will not impact image quality so long as it is operating within bandwidth and error rate performance specification parameters.*

Sound Fidelity—Speech Quality   References to sound fidelity include terms like *FM Quality* and *CD Quality*. Assessing sound quality tends to be almost totally subjective. However, the ability of a transmission channel or other form of media to carry sound can be measured and expressed objectively in terms of frequency response, signal-to-noise ratio, distortion, and, in the case of stereo or multiple channel sound transmission, separation.

Speech quality is a telephone term usually expressed in terms of compliance with a particular standard. The most commonly used reference is ITU Rec. G.711, defining a toll quality speech codec. Adopted many years ago, it is, and has been, the benchmark for speech quality in international telephony that most succeeding codecs have been measured against. Essentially, it relies on an approach where the speech signal is sampled at an 8Khz rate and quantized to 8-bits. The output of the encoder is a 64Kbs stream, matching the 64Kbs channel rate provided by the public switched telephone network. G.711 includes two encoding laws. The Mu-Law version is used in Japan and North America. The A-Law version is used in the rest of the world.

ITU Rec. J.52 covers transmission of high-quality sound-program signals using one, two, or three 64 Kbs channels per mono signal, and/or up to six per stereo signal. The system described in J.52 enables transport of sound signals using new coding techniques included in ISO11172-3.

Service   Service is the right to use network equipment and facilities. Many people confuse services and facilities. Still others say product when they should say service, and still others confuse technology and methodology with services. For example, a DS3 or E3 is a facility, not a service or product. Telephone companies provide services, not products. The only products they *make* are phone bills. ISDN, ATM, and Frame Relay are often called a service, but in reality they are simply a technology or methodology. For example, ISDN is an access method; ATM and Frame Relay are transport methods or technologies.

Standards   Standards are documents containing processes, guidelines, and recommended practices established accredited standards bodies such as AES, ANSI, ETSI, IEEE, IETF, ITU, MPEG, and SMPTE. The use of standards is voluntary, not mandatory, although they can and are often used as references in contracts and other legal documents.

Quite often, a Special Interest Group (SIG) or their work output is mistakenly referred to as a standard. The ATM Forum, Video

Services Forum, and SONET Interoperability Group are all special interest groups dedicated to promoting and advancing their collective and common interests. SIGs typically publish documents and support development of standards, many of which would not likely exist without SIG effort and resources. Less common is use of the term *de-facto standard*, which is simply another way of referring to a proprietary product or service.

SONET Synchronous Optical Network

SDH Synchronous Digital Hierarchy

SDN Software Defined Network

SNMP Simple Network Management Protocol

Stratum Clock The stratum clock is a network clock reference originally arranged from 1 through 5 levels, in which level 1 is the most precise level and 5 is the least precise level.

Transmission Channel A fixed amount of bandwidth usually expressed in megabits per second is called a *transmission channel*. For example, a DTV Transmission channel runs at a precise rate of 19.392685Mbs. A T1 runs at 1.544Mbs and a DS3 at 44.736Mbs.

TCP Transmission Control Protocol

TCP/IP Transmission Control Protocol (over) IP (Transport)

VPN Virtual Private Network

BIBLIOGRAPHY

Author Unknown. *Digital Satellite Technology*. Revision 1, Washington, DC: INTEL-SAT, 1992.

Author Unknown. *The Book: More Engineering Guidance for the Digital Transition*. NVISION, Inc., 1999.

Brinkley, Joel. *Defining Vision: The Battle for the Future of Television*. New York: Harcourt Brace, 1997.

Cooke, Nelson M. *Mathematics for Electricians and Radiomen*. New York: McGraw-Hill, 1942.

Dutcher, Bill. *Managing IP Addresses*. New York: Wiley Computer Publishing, 2000.

Finkler, Stephen. *Finance & Accounting for Non-Financial Managers*. Third Edition, New York: Prentice-Hall, 2002.

Garbin, David A. and Pecar, Joseph A. *Telecom Factbook*. Second Edition, New York: McGraw-Hill, 2000.

Goralski, Walter. *SONET/SDH*. Third Edition, Berkley, CA: McGraw-Hill/Osborne, 2002.

Hall, Eric A. *Internet Core Protocols: The Definitive Guide*. Sebastopol, CA: O'Reilly & Associates, 2000.

Hull, Scott, et al. *Content Delivery Networks: Web Switching for Security, Availability, and Speed*. New York: McGraw-Hill/Osborne, 2002.

Kantor, Rosabeth Moss. *The Change Masters: Corporate Entrepreneurs At Work*. London: George Allen & Unwin, 1983.

Kahn, Daniel, Buchanan, Frank, Miller, Mark, Terjian, David, Pasek, Craig, Sword, Duane, Sitavi, Raul, and Lowe, Ronald. 1997 Broadband Network I&M Test Seminar, Palo Alto, CA: Hewlett Packard, Measurements Division (Since renamed "Agilent"), 1997.

Inglis, Andrew and Luther, Arch C. *Video Engineering*. First and Second Editions, New York: McGraw-Hill, 1993, 1996.

Iseminger, David. *Windows 2000 Quality Of Service*. New York: Macmillan, 2000.

Luther, Arch C. *Video Camera Technology*. Boston: Artech House, 1998.

McConkey, Dale D. *Planning Next Year's Profits.* American Management Association, Inc., 1968.

McDysan, David. QOS & Traffic Management. In *ATM and IP Networks.* New York: McGraw-Hill, 2000.

McDysan, David and Paw, David. *ATM & MPLS Theory & Application: Foundations of Multi-Service Networking.* New York: McGraw-Hill, 2002.

Miller, Mark A. *Analyzing Broadband Networks.* Third Edition, New York: McGraw-Hill, 2000.

Miller, Mark A. *Troubleshooting TCP/IP.* Third Edition, Foster City, CA: M&T Books, 1999.

Moore, Geoffrey A. *Crossing The Chasm: Marketing and Selling High-Tech Products to Mainstream Customers.* New York: HarperCollins, 1995.

Moore, Geoffrey A. *Inside the Tornado: Marketing Strategies From Silicon Valley's Cutting Edge.* New York: HarperCollins, 1995.

Naisbitt, John and Aburdene, Patricia. *Re-inventing the Corporation: Transforming Your Job and Your Company for the Information Society.* New York: Warner Books, 1995.

Robin, Michael. *Television in Transition: A Compendium of Articles Published in Broadcast Engineering.* Quebec, Canada: Miranda Technologies, Inc., 2001.

Schwartz, Evan L. *The Boy Genius and The Mogul* (also published under the title The *Last Lone Inventor: A Tale of Genius, Deceit, And The Birth Of Television*) Perennial, an imprint of HarperCollins, 2002.

Senge, Peter, Kleiner, Art, Roberts, Charlotte, Ross, Richard, Roth, George, and Smith, Bryan. *The Dance of Change: The Challenges of Sustaining Momentum in Learning Organizations.* New York: Doubleday, 1999.

Simoneau, Paul. *SNMP Network Management.* New York: McGraw-Hill, 1999.

Sinnreich, Henry and Johnston, Alan B. *Internet Communications Using SIP: Delivering VoIP and Multimedia Services with Session Initiation Protocol.* New York: John Wiley & Sons, 2001.

Slurzberg, Morris and Osterheld, William. *Essentials of Electricity for Radio and Television.* Second Edition, New York: McGraw-Hill, 1950.

Stallings, William et al. *Handbook of Computer Communications Standards.* Volume 2 & 3, Carmel, IN: Macmillan, Inc., 1991.

Stevens, Mark. *The Big Eight: An Incisive Look Behind the Pinstripe Curtain of the Eight Accounting Firms Whose Practices Affect the Lives and Pocketbooks of Every American.* New York: Collier Books, 1984.

Verma, Pramode K. et al. *ISDN Systems, Architecture, Technology, and Applications.* Englewood Cliffs, NJ: Prentice Hall, 1990.

Westman, H. P. et al. *Reference Data for Radio Engineers.* Fourth Edition, New York: International Telephone and Telegraph Company, 1964.

Wolf, Michael J. *The Entertainment Economy: How Mega-Media Forces Are Transforming our Lives.* New York: Times Books Random House, 1999.

Wiese, Michael and Simon, Deke. *Film & Video Budgets.* Second Edition, Studio City, CA: Michael Wiese Productions, 1997.

REFERENCES

Audio Engineering Society, Inc. *AES10-1991 Recommended Practice for Digital Audio Engineering—Serial Multi-channel Audio Digital Interface (MADI)*.

Audio Engineering Society, Inc. *AES information document for digital audio engineering 1996, Reaffirmed 2001—Guidelines for the use of the AES3 interface*.

Author Unknown. *Dictionary of Internetworking Terms and Acronyms*. Indianapolis: IN: Cisco Press, 2001.

Brown, Charles L. "The Bell System" From *Encyclopedia of Telecommunications*. Marcel Dekker, Inc., 1999.

Carpenter, Brian. *Internet Architecture Board (IAB) Architectural Principles of the Internet*. Internet Engineering Task Force (IETF) Request For Comments (RFC), 1968. http://isi.edu/in-notes/rfc1958.txt

Clarke, Arthur C. *Exploration of Space*. New York: Pocket Books, 1979.

Gigabit Ethernet Association White paper. *Gigabit Ethernet Over 4-Pair 100 Ohm Category 5 Cabling*, http://www.10gea.org/GEA1000BASET1197_rev-wp.pdf

Goralski, Walter. *SONET/SDH*. Third Edition, New York: McGraw-Hill, 2002.

Held, Gilbert; *Dictionary of Communications Technology*. Second Edition, New York: John Wiley & Sons, 1995.

http://www.lsi.usp.br/rbianchi/clarke/ACC.ETRelaysFull.html

IEEE Standards, http://standards.ieee.org/

Lammle, Todd and Hales, Kevin. *CCNP™ Switching Study Guide*. San Francisco: Sybex, 2001.

Luther, Arch C. *Digital Audio and Video Systems*. Boston: Artech, 1997.

Moore, Gordan. A phrase referring to doubling computer power for a given fixed cost every 18 months.

Newton, Harry. *Newton's Telecom Dictionary*. 18th Updated and Expanded Edition, Gilroy, CA: CMP Books, 2000.

OSI/IEC 10731 http://www.OSI.org/OSI/en/OSIOnline.frontpage

Park, David G. Jr. *Good Connections: A Century of Service by the Men and Women of Southwestern Bell*. First Edition, Southwestern Bell Telephone Company, 1984.

RFC0894 Transmission of IP Datagrams over Ethernet Networks, http://www.ietf.org

Robin, Michael and Poulin, Michel. *Digital Television Fundamentals.* New York: McGraw Hill, 1997.

Saltzer, J. J., Reed, D. P., and Clark, D. D. "End-to-End Arguments in System Design." *ACM TOCS*, 2(4):277-288, 1984.

Sennreich, Henry and Johnston, Alan B. *Internet Communications Using SIP.* New York: John Wiley & Sons, 2001.

SMPTE Standards, http://www.smpte.org/smpte_store/standards/

Telarc International. *Telarc Revisits the "Blast from the Past" with New DSD Technology, Newly Recorded Cannons, Carillons and Choruses* http://www.telarc.com/gscripts/title.asp?gsku=0541

Whalen, David J. *Communications Satellites: Making the Global Village Possible* http://www.hq.nasa.gov/office/pao/History/satcomhistory.html

Milton Keynes UK
Ingram Content Group UK Ltd.
UKHW051847071024
449327UK00025B/1886